T0295557

Environmental Social Governance

Environmental Social Governance
Managing Risk and Expectations

Karlheinz Spitz, John Trudinger, and Matthew Orr

CRC Press
Taylor & Francis Group
Boca Raton   London   New York   Leiden

CRC Press is an imprint of the
Taylor & Francis Group, an **informa** business

A BALKEMA BOOK

Book layout and design: Diky Halim & Wynne
Taletha (www.did-design.com)

Cover photo: GermanL/Shutterstock

First published 2022
by CRC Press/Balkema
Schipholweg 107C, 2316 XC Leiden, The Netherlands
e-mail: enquiries@taylorandfrancis.com
www.routledge.com – www.taylorandfrancis.com

CRC Press/Balkema is an imprint of the Taylor &
Francis Group, an informa business

Library of Congress Cataloging-in-Publication Data

A catalog record has been requested for this book

ISBN: 978-0-367-68055-8 (hbk)
ISBN: 978-0-367-68056-5 (pbk)
ISBN: 978-1-003-13400-8 (ebk)

DOI: 10.1201/9781003134008

Typeset in Times New Roman
by codeMantra

Contents

Foreword

As renowned ethicist Robert B. Reich[1] observed "...individuals, not corporations, must be held accountable for undermining the common good." Environmental damage is not caused by corporations; it is the result of actions, or lack of action, of individuals. Like it or not, executives and Directors are responsible and accountable for their own actions and inactions, and also to a large extent, for those of their employees. It follows that industry leaders need knowledge and information that will assist them in discharging these responsibilities. The purpose of this book is not to convert executives and Directors to become environmentalists. Rather, it is to identify and explain the various environmental and social risks that leaders of industrial companies are likely to encounter and to suggest some strategies for responding to the challenges that will arise. In particular, it is hoped that this book will equip business leaders to ask the appropriate questions of the relevant staff, to seek access to external expertise, and to assess the adequacy of the answers they receive.

Two of this book's authors – Karlheinz Spitz and John Trudinger – have enjoyed long careers as environmental consultants which provided the opportunity to observe scores of industrial companies, small and large, operating in a variety of jurisdictions throughout the world. When, following the success of their book[2] *Mining and the Environment – From Ore to Metal*, they decided to produce a book aimed at senior executives and Directors, they realised that their perspective – as outsiders – needed to be augmented

---

[1] Reich, R. B. (2018). *The Common Good*, (Alfred A Knopf, New York).

[2] Spitz, K. and Trudinger, J. P. (2019). *Mining and the Environment – From Ore to Metal*, 2nd ed., (CRC Press – Taylor & Francis Group, London, UK)

by the views of someone with corporate experience as an insider. Matthew Orr, with experience in both large multinational corporations and smaller local companies, fulfils this role.

Like most people, we have experienced many instances of frustration in our professional life. It seems every 5 years or so, a new vision of rectitude proclaims a "new" environmental paradigm with another catchy name and still another three-letter abbreviation. To our ears, these catchy names sound like argie-bargie, a word salad of warmed-over MBA and consultant-speak: sustainability... offsets... Free Prior Informed Consent or FPIC... metrics... Environmental, Social and Governance or ESG... community involvement... corporate responsibility..., and the list goes on. These words have entered the business jargon and are so overused or misused that they may no longer have real meaning. Hopefully, this book will serve as a guide through the jargon to the issues and principles at stake.

We also notice over the last half-century, at least within the OECD, the doctrine of maximising shareholder value has been the definitive tool for measuring the performance of executives of public corporations and the companies they lead. While even Jack Welch himself – the idea's leading exponent – by 2009 had come to call it "the dumbest idea in the world," many business leaders continue to embrace this doctrine as the main measure of business success. As is becoming clear entering the third decade of this millennium, this is changing rapidly and not necessarily voluntarily. In a 2019 Statement about the purpose of corporations, the Business Roundtable, representing chief executives of nearly 200 leading U.S. companies including Apple, Pepsi, and Walmart, said business leaders should commit to balancing the needs of shareholders with customers, employees, suppliers, and local communities. This statement made headlines: "Shareholder Value Is No Longer Everything, Top CEOs Say" *New York Times* (19 August 2019); "Move Over, Shareholders: Top CEOs Say Companies Have Obligations to Society" read the front page of the *Wall Street Journal* on the same day.

Since 1978, the Business Roundtable has periodically issued "Principles of Corporate Governance." Each version of the document issued since 1997 endorsed principles of shareholder primacy – that corporations exist principally to serve shareholders. The 2019 Statement is a shift toward a new standard for corporate responsibility. This book is concerned with the environmental and social aspects of corporate governance and does not address other governance aspects.

Maximising one business value to the exclusion of all others is no longer acceptable. Neither is it sustainable. With evolving environmental and social issues on the horizon – think climate-proofing, gender equality, sustainability, Equator Principles, and biodiversity conservation, all part of Environment, Social, and Governance or ESG – the coming years are shaping up to be action-packed for corporate Boards.

A Bloomberg analysis of quarterly earnings calls and other conference calls related to the 23 members of the S&P 500 Energy Index showed that the mention of ESG and other sustainability-related terms soared in the first quarter of 2021 (compared to the same quarter the previous year). The use of ESG-related terms among oil and gas companies jumped from 36 in the first quarter of 2020 to almost 300 in the first quarter of 2021 (www.environmentalleader. com). While it may take years to unravel what the advent of ESG factors will do to business, the hyperdrive ESG shift will probably leave many business leaders struggling to achieve and maintain the balance between profit and ESG – their responsibility to shareholders versus to society.

We have a genuine intention of sharing the best of our years of experience, understanding, knowledge, and insights to help Board members respond to these environmental and social business changes. We also live in a world today where one disgruntled person, one small issue, or a public misconception can override all practical and economic common sense to cause years of delay in project developments, increase costs to unsupportable levels, or produce insurmountable political barriers. Challenges posed and opportunities created by the shift toward greater environmental awareness and sustainability present the industry with complicated and seemingly intractable social and environmental risks.

This book offers high-level guidance to senior management to manage and communicate material risks in the context of the Company's business model, sustainability impacts, and stakeholder relationships. The text focuses on addressing environmental and social risks from the viewpoint of senior management and non-executive Board members, sidestepping technical details and the gobbledygook all too often associated with academic writing. Board members need useful information quickly and will not suffer a lengthy, wandering approach. The discussions will assist management in identifying "fatal flaws" if and as they arise.

As industries are sourcing more and more production from developing countries with different priorities and expectations than those of the developed world, challenges confronting these operations receive special attention in this book. While the environmental and social issues are similar between developed and developing countries, we recognise that corporate governance as practiced in OECD countries is not recognised in some jurisdictions and the equivalent of Board oversight may not apply in some countries. Nevertheless, the principles and approaches discussed in this book are relevant to industrial developments wherever they take place.

Our text offers a taste of some of the more critical environmental and social challenges Board members are likely to face (emphasising the mining sector given the authors' professional expertise). It may help avoid playing whack-a-mole with ESG risks and mitigate challenges before they spiral out of control.

About the authors

Karlheinz Spitz, holding dual degrees in Business Administration and Civil Engineering, is an environmental consultant of international repute. He is the lead author of several books, including *Mining and the Environment – From Ore to Metal*, published by Taylor & Francis (2nd edition), and co-author of the acclaimed coffee table book *The World of Mining*. His primary interest is the environmental assessment of resource development projects in developing countries, covering a wide range of natural resources and diverse spectrums of social and ecological settings.

He regularly advises lenders on Equator Principles aspects in their investments and offers ESG training to business leaders. After working in senior positions for international consultancies, Karlheinz founded Greencorp, an environmental consultancy providing advice to clients who face a myriad of environmental laws and regulations, combined with an increasingly complex web of ESG risks and expectations.

This catchphrase *We provide solutions where others see problems* captures his business philosophy, prompting him and his co-authors to write this book.

John Trudinger is an environmental consultant with more than 50 years of project experience. Qualified as a geologist, his initial experience involved geotechnical investigations for large water supply and transportation projects. Since the early 1970s, he has provided environmental expertise to a variety of resource development, power generation, and infrastructure projects throughout Australia, North America, and Asia in a career that has spanned the entire environmental movement to date. He is co-author of *Mining and the Environment – From Ore to Metal*, published by

Taylor & Francis (2nd edition), as well as the coffee table book *The World of Mining*. John has served as Director in a NYSE publicly listed Company and several private companies.

Matthew Orr has over 30 years of experience in managing environmental and social risks associated with mining operations and related stakeholder engagement in both operational and corporate roles. This was largely spent in developing countries and at challenging sites characterised by high rainfall, high seismicity, high-value ecosystems, and nearby rural communities. His original qualification was in forest science, and his technical expertise includes project impact assessment, site rehabilitation, biodiversity, acid mine drainage, site water management, tailings disposal, occupational safety, training systems, crisis management, project approvals, enterprise risk assessment, sustainability reporting, and management systems. He has successfully led the design and implementation of safety management systems for several large mining companies. Mat has contributed to the development of industry-leading practices in managing environmental and social impacts and has managed to document some of these in 14 conference papers.

His career highlight was gaining Company support for hosting and organising the inaugural Australian Acid Mine Workshop in 1993. He now lives and works in Indonesia.

Acknowledgements

Having an idea and turning it into a book is as hard as it sounds. It was made much easier by the support of Robert McDonough, our friend and colleague, for many years. From reading and providing inputs and advice on early drafts to proofreading the final text, his contributions were invaluable.

Special thanks to Alistair Bright, the ever-patient Publishing Manager, and Diky Halim, the most incredible designer we could ever imagine. Complete thanks to Ed Gleeson, who made this book a better text by sharing his insights and thoughts. Thanks to the Board of PT Agincourt Resources for allowing use of Company material that may support continual improvement in safety and risk management in other companies.

"Yes, the planet got destroyed. But for a beautiful moment in time we created a lot of value for shareholders."

Source: Tom Toro/CartoonStock

Environmental and social aspects of governance in a changing environment

The Board's role

Be in no doubt, Environmental, Social and Governance (ESG) factors have become increasingly important to business leaders responding to today's ever-changing business environment. Ignoring this trend would indicate that we, as business leaders, have become unmoored, not just from our duty, but from reality itself.

There is pressure *from below* through societal movements and evolving consumer behaviours; *from within*, driven by activist shareholders and institutional investors, and by socially conscious employees; and legislative pressure *from above* due to incremental advances of ESG regulations at national and international levels. All are emblematic of the whipsaw ride today's corporate business leaders themselves are experiencing.

The pressure from within applied by shareholders to the Rio Tinto Board in 2020 following the destruction of the archaeologically and culturally rich Juukan Gorge caves in Western Australia for an iron ore development illustrates the magnitude and consequences of ESG mistakes. This resulted for Rio Tinto, previously regarded as an industry leader, being identified as the sixth most distrusted brand in Australia in 2021. There was also pressure from above, as the Chief Executive Officer (CEO) was summoned to appear before the Australian Parliament inquiry into the destruction of the caves.

Rio Tinto's first attempt at contrition – cutting several millions from the bonuses of three key executives – was a sentence perceived as way too light to reflect appropriate accountability. Shareholders asked for heads to roll, eventually leading to the CEO and other senior Company executives' departure. And it did not matter that senior executives in the firing line claimed that they were not aware of the Juukan Gorge caves' significance. As corporate business

DOI: 10.1201/9781003134008-1

Walk the Line (Randy Langstraat)

leaders, they were held accountable. To quote Rio Tinto chairman Simon Thompson[1] (National Resources Review, 11 September 2020): "We have listened to our stakeholders' concerns that a lack of individual accountability undermines the group's ability to re-build that trust and to move forward to implement the changes identified in the Board review."

Assailed from above and below, companies with high ESG performance are proven to have lower risks, higher returns, and more resilience in times of crisis. Businesses with leading ESG performance also attain superior access to capital. It is easy to see why Boards must understand how their risk oversight role specifically applies to ESG-related risks.

Not so long ago, the Board's overarching duty was to maximise profits for the benefit of shareholders – the "bottom line." Now, however, as the above example illustrates, at least three "bottom lines"

---

1 The chairman announced on 3 March 2021 he would not seek reelection post 2022, stating: "I am ultimately accountable."

require attention in the boardroom. And, in addition to the Triple Bottom Line (TBL, see BP 02), modern Directors must consider Sustainability, Corporate Social Responsibility (CSR), the Company's Social Licence to Operate (SLO), and, possibly, Enlightened Shareholder Value (ESV). It seems every 5 years or so, a new vision of rectitude proclaims a "new" paradigm with another catchy name and still another three-letter abbreviation, and the list continues to grow. To a large extent, these all represent the same concepts: environmental protection, human rights, and support for communities.

ESG CONSIDERATIONS NO LONGER "NICE TO HAVE" BUT A MUST

As acknowledged by the 2018 World Economic Forum, eight of the top ten global risks are ESG related (The Global Risks Report 2018). ESG, a term first coined in 2004 in a study by the UN Global Compact entitled "Who Cares Wins" is a broad set of concepts with no standard definition. However, we can generally agree on the scope of issues it covers – how our Company performs as a steward of our environment, how our Company manages relationships where it operates, and, ultimately, how ethically it (and we) acts. Unsurprisingly, ESG is quickly rising to the top of Board agendas. So, what are the ESG issues to be addressed by the Board?

> When you leave the light on,
>   you're not the only one who pays.
>
> – WWF

The most critical issue, of course, is to ensure regulatory compliance of operations in ESG matters. Occasionally we may be asked questions such as: What is the value associated with an environmental compliance programme, or what are the potential costs of noncompliance? Inquiries like this are missing the fundamental point: To operate legally, the Company needs to comply with prevailing laws and regulations. To run the business knowingly in regulatory noncompliance exposes the Company, the Board, and responsible Company officials to legal risks.

Regulatory compliance issues aside, a highly essential but frequently misunderstood topic is sustainability, further discussed in BP 02. It might not be the leading boardroom topic, but sustainability is now central to corporate competitiveness and a Company's continued ability to operate.

In its broadest sense, sustainability embodies most if not all the concepts of the TBL, CSR, and SLO with additional emphasis on ecological values, notably biodiversity (discussed in BP 10).

Climate change (discussed in BP 13) is another prominent issue. As underlined by BlackRock CEO Larry Fink in his 2020 "Letter to CEOs," climate change is fundamentally reshaping finance as climate risk is now generally acknowledged as an investment risk (Fink 2020).

The Equator Principles (discussed in BP 04) represent a risk management framework adopted by financial institutions to determine, assess, and manage environmental and social risk in capital projects. Their primary intent is to provide a minimum standard for due diligence and monitoring responsible risk decision-making. The "Bettercoal" Code was launched in 2013 to assess, assure, and sustain stringent ethical, environmental, and social performance in the coal mining supply chain (admittedly a tall order, critics may say). The "Copper Mark" of 2020 is a global standard to ensure responsible production and trading of copper, inspired by the United Nations Sustainable Development Goals. As with other Sustainability Development Frameworks, these codes aim to help members understand, manage, and mitigate ESG risks in their operations and supply chain.

ESG ENTERS THE BOARDROOM

ESG has changed the *raison d'être* for the Board: "Doing Things Right is of No Value if You're Not Doing the Right Thing(s)." Simplistically, this sums up the Board's role: The Board's fiduciary duty is to oversee a Company's strategy, risk, and capital allocation. In his seminal book – *Corporate Governance: Principles, Policies, and Practices* – Bob Tricker (Tricker 2015) identifies and discusses the four primary Board responsibilities and associated activities: strategy formulation, policymaking, monitoring and supervising, and providing accountability. These are depicted as a framework matrix below. All four responsibilities/activities include ESG considerations.

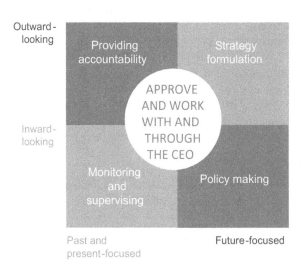

Strategy formulation includes environmental, safety, and social issues such as stakeholder engagement, risk assessment and management, community development, and emergency preparedness and response. The Board usually works with management to spot ESG risks (e.g., greenhouse gas emissions and ecosystem degradation) as well as opportunities (e.g., energy efficiency and community development initiatives). Collaboration enables the Board to integrate values, goals, and metrics into business strategies to mitigate ESG risks and inform decisions aligned with their fiduciary duties, thus avoiding eleventh-hour belated attempts to embrace emerging ESG factors.

Policymaking is an essential task of management, with the Board responsible for ensuring the adequacy and appropriateness of policies. Policies will be informed by Government regulations, project approval conditions, guidelines promulgated by industry groups, and internal policies that are directed at specific Company circumstances. Environmental policies will address matters such as greenhouse gas emissions, pollution control, energy consumption, waste minimisation and disposal, and biodiversity. Social policies may include stakeholder engagement, land acquisition, employment and remuneration, anti-corruption, grievance identification and resolution, and support for local businesses and community activities.

Monitoring and supervising are required to assess compliance with external and internal policies. The extent of monitoring programmes, the parameters to be measured, and their frequencies will depend on the nature of the industry, its area of influence, and its potential environmental impacts. The Board needs sound knowledge of Company operations to guide what questions to ask and where and how to obtain reliable information. As with financial audits, environmental and social reviews, whether internal or external, are vital sources of information that must be part of a continuing programme.

Accountability involves responding to external stakeholders about overriding activities and performance while ensuring profitability and transparency. Errors or weaknesses in the other three responsibilities affect the degree of stress or pain associated with accountability.

BOARDS THAT ADAPT WILL PROSPER

Few principles guide Boards to strengthen Company resilience to ESG-related risks. First, as business leaders, we must ensure that Board members have access to ESG skills, imparting ESG training to Directors as necessary. Given the ever-changing business environment, auditing the ESG skills of Directors is not a one-time exercise but should be done regularly, e.g., as part of the Board evaluation process. Second, the Board should judge which ESG risks the Company is most exposed to, and as necessary seek specialist advice on how to mitigate risks. Third, the Board should evaluate whether the monitoring of ESG risks is adequately resourced and delegated to management. Finally, the Board needs to assess ESG compliance at least annually and encourage high-quality ESG reporting, a significant element of proactive engagement with all stakeholders on ESG-related issues.

Management runs the business;
the Board ensures that it is being well run and run in the
right direction.

(Tricker, op cit)

In discharging their duties, Company Directors need to be aware
of various statutory, administrative, and other legally binding requirements. Then, depending on the nature of the business, there
may be industry codes that members of industry associations are
expected (or have pledged) to observe. More difficult to address
are the expectations of stakeholders; the wider the stakeholder diversity, the more likely that there will be conflicting expectations
among different stakeholder groups. And, most difficult of all may
be the political implications of Company decisions and actions.
This particularly applies to environmental and social issues, which
inevitably have political overtones.

Each Company also needs to decide its attitude toward and the
degree to which it will engage with Non-governmental Organisations (NGOs); the combined wisdom and experience of Directors
will be valuable in this decision. Partnerships with civil society
groups are challenging due to their asymmetric nature, particularly given NGOs' complex triple role as agents of change, pressure
groups, and fundraising entities.

A Company's policy objectives are usually addressed by employing Standard Operating Procedures (SOPs), of which there
may be dozens or even hundreds for any particular operation.
Formulation of SOPs is not the Board's role, other than the Board
should ensure that there are procedures in place to meet all essential policy objectives. However, setting up an "oxygen tent" to
foster ESG considerations in management decisions is an important role.

BOARD COMPOSITION

According to the Corporate Governance Principles and Recommendations of the Australian Securities Exchange, "A listed entity
should have a board of appropriate size, composition, skills and
commitment to enable it to discharge its duties effectively" (Corporate Governance Council 2014). Traditionally, Company Directors have been selected based on their business expertise and

experience; the combined contributions of all Board members are intended to positively guide the Company and help protect it from adverse outcomes. Experience in the particular industry is beneficial, if not essential, but diversity among non-executive Board members is becoming increasingly valued.

The typical boardroom has members with proven business success plus financial, marketing, and legal credentials. Also commonly appointed are Directors with political influence and experience in labour relations. However, it is unusual for a Board to include a Director with a sound knowledge of environmental and social issues. Given the increasing importance of social and ecological matters, particularly in the resource extraction and processing industries, this is somewhat surprising. Most Boards, especially of larger organisations, would benefit from the inclusion of a Director with ESG credentials. Domain-specific knowledge and relationships are as relevant for those areas as for any others.

THE COMING IMPACT OF ESG ON STRATEGY

Sustainability has become a blanket term – a catch-all for any Company's efforts to "do good." ESG, on the other hand, pinpoints three specific elements that are crucial to today's business leaders and investors. Institutional investors, notably State Street and BlackRock, have stressed the importance of disclosing comparable and meaningful ESG metrics by their portfolio companies.

So what's next? Increasingly, stakeholders expect companies to release high-quality, accurate ESG data suitable for investment decisions. To achieve "Investment Grade" ESG status, companies must collect timely, accurate, complete, and auditable data. With this development, the question isn't "Can you collect ESG data?" but rather, "How trustworthy are the data you're collecting?"

The market will determine the exact system by which companies will report ESG data (BP 02 elaborates on the Global Reporting Initiative Standards, the first global sustainability reporting standards). Still, it is safe to assume that from now on, a Company that is to be seen as a valuable member of society must disclose significant amounts of quantifiable information that will permit comparisons within and across industries. What types of data this will involve in the post-COVID-19 era, at this writing, cannot be reliably predicted.

There is also a real prospect of future product differentiation (and price premia) for sustainably produced metals and industrial goods. Just like ethically produced coffee costs more, conscientious Tesla drivers might like to know where the copper in their cars' engines came from and pay a bit more for the privilege. The millennials will drive higher demand for sustainably sourced goods, demonstrated through increased transparency of supply chains. The general population is becoming increasingly concerned about what goes into everyday goods and services, and the footprint they leave on this Earth. In 2018, Apple became the first Company to map its supply chain from manufacturing to the smelter for tin, tantalum, tungsten, cobalt, and gold, considering conflict, human rights, and other risks, going above and beyond what is required by law. Consumers care, and they speak with their wallets.

Premium
Pricing

DIRECTORS' DILEMMA: ESG OBLIGATIONS OR OVERKILL IN THE BOARDROOM

Driven by well-publicised natural disasters such as the devastating forest fires raging across parts of Indonesia in 2015 or Australia's bushfire crisis in 2019–2020, and by campaigns led by high-profile individuals such as Al Gore and Greta Thunberg, ESG values will continue to assume ever greater importance at Board level. Rightly or wrongly, some Directors consider that ESG has already become overblown, too much of a good thing. Be that as it may, ESG has been a big focus with institutional investors in recent years, and the Board needs to live up to investors' expectations. ESG considerations are unlikely to diminish in the foreseeable future.

Environmental activists have become increasingly successful at leveraging influence and fostering outrage to achieve their aims, often prepared to exaggerate and even lie to achieve their goals. Various campaigns have been implemented, including:

- Persuading financial institutions to avoid funding coal projects. This has been so successful that it is now being extended to other fossil fuels;
- Pressuring individual companies to resign from industry associations that do not subscribe to the green agenda; and
- Advocating that major companies become carbon neutral.

Furthermore, companies are expected to track and report Scope 3 emissions, which are emissions generated by customers.

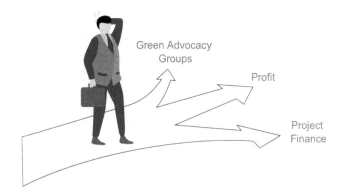

Green advocacy groups may use a small stock position to have an agenda item included at a stockholders' meeting. They seek and sometimes receive support from large investment groups such as pension funds. However, individual stockholders are not involved, and it is challenging to assess investors' views accurately. Whatever the outcome of a campaign, the media can be relied on to cover the issue, including "naming and shaming," so the green agenda is advanced even when the vote is lost.

Industry associations draw their membership and funding from many companies with ranges of circumstances and viewpoints. Accordingly, such associations are not in a position to subscribe to all elements of the green agenda. Many companies assert that membership of an industry association has many benefits, including in social and environmental matters, and rather than resign and forego these benefits, it is better to retain membership, and in so doing help the association better serve its members. The inclusion of Scope 3 emissions in carbon footprint calculations is even more difficult for companies seeking to become carbon neutral. If a Company somehow achieves carbon neutrality for its own emissions plus those of its customers, does that then absolve customers from dealing with their own carbon footprints? These and similar considerations provide major dilemmas for Directors. Although they may be favourably disposed toward improving environmental and social performance, they are affronted by the tactics and troubled by the disruptions to operations and the extra public relations efforts required to meet the challenges as they arise. So, discussions in the boardroom can become exceptionally difficult; individual Directors may be strongly opposed to a motion, on principle. However, the question is not what the Directors feel, but what is best for the Company and its stakeholders. An honest appraisal would have to acknowledge which ESG factors deserve prominent Board attention. And, of course, what is favourable to one "bottom line" may be detrimental to another.

LOOKING AHEAD: IDENTIFYING KEY TRENDS IN ESG

At the heart of the Board's role is to provide management with forward-looking strategic guidance. It calls for the Board to continually assess opportunities and risks in a dynamic and uncertain political, social, and business environment and to accept the mantle of responsible ESG leadership.

Toward the end of the 1990s, Nike products became associated with slave wages, forced overtime, and arbitrary abuse (Locke 2003). The sportswear giant was suffering under a rising tide of scandals. It was attacked by a new form of activism that targeted consumer-facing brands to curb environmental and labour abuse further up the supply chain. In an apparent change of strategy, Nike recruited a new Director of corporate responsibility to transform Nike from a Company that was synonymous with sweatshops to a recognised sustainability leader. This forward-looking strategic thinking, though reacting to reputation damage, led to a significantly burnished image for Nike, no small matter for a firm selling high-priced sporting goods to consumers who tend to be conscious of their own images.

Active ESG management with forward-looking sustainability initiatives is a tool for future-proofing the Company in a business environment where the industry is under pressure to cut emissions, demonstrate a commitment to communities, and address global challenges such as biodiversity conservation. Here, the long and short of it is that business leaders set the tone for their companies, recognising that the real benefits from active ESG management will take time to make themselves felt.

Future ESG trends are discussed in the closing chapter of this book, but it is appropriate at this stage to draw attention to one emerging trend – big data. Can big data have a big impact on sustainability? In recent years, data gathering, computing power, and connectivity have advanced beyond imagination. We now have more information than ever before at our fingertips. In 2017 alone, society generated more data than in the previous 5,000 years, less due to changes in activities than to the burgeoning ability to record them and store records. We are arguably just starting to scratch the surface of how businesses can use all this extra information to become more sustainable. One advantage of big data is that it unlocks companies' ability to understand and act on what may often be their most significant environmental impacts – those outside their direct control. Big data helps companies to be conscious of direct effects and those produced throughout their entire value chain.

Green data is the application of big data to curb global warming. Green data can contribute to creating more sustainable companies. Data allow energy management, resource use, and plant maintenance to be optimised, while emissions from production and transportation can be reduced.

In the extractive industries, understanding the host environment is vital for the operator to foresee and monitor potential impacts on air, water, human health, and biodiversity. The sector must also respond to implications of a changing climate and other anthropogenic pressures, e.g., forest clearing in tenements unrelated to project activities. Big data offers a wide range of novel techniques to support a deeper understanding of the host environment in all its complexities (e.g., land use change, forest loss, surface water temperature, or even changes in seabed or sightings of endangered species and migrating birds). They also support developing well-grounded environmental mitigation and adaptation strategies. One big data application with many beneficial uses that we are all familiar with is Google Earth.

2001–2010

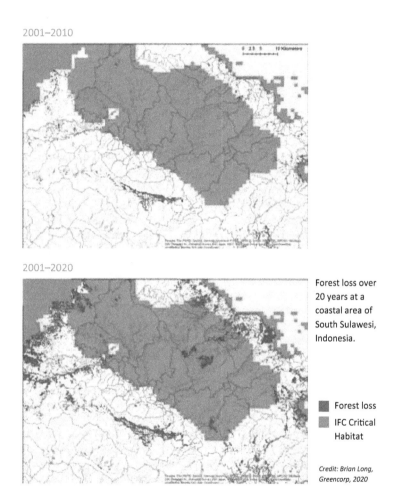

2001–2020

Forest loss over
20 years at a
coastal area of
South Sulawesi,
Indonesia.

◼ Forest loss

◼ IFC Critical
   Habitat

*Credit: Brian Long,
Greencorp, 2020*

Forest loss over 20 years at a coastal area of South Sulawesi, Indonesia. (Credit: Brian Long, Greencorp 2020.)

Yes, sustainability can be a strategy

Sustainability is almost certainly the most widely used environmental "buzzword" of the past decade. It has been commonly used and misused to denote a variety of concepts, as it defies easy explanation. These days, despite some confusion over its meaning, regulators and industries, including mining, have embraced the overall premise, as environmental sustainability can be considered a human right. It has become the "new normal" for business leaders to embrace sustainability and to speak out on sensitive social and environmental issues, apart from the Company's financial performance.

CONCEPT OF SUSTAINABLE DEVELOPMENT

In 1987, the United Nations released "Our Common Future," commonly called the Brundtland Report. This established what remains the most widely quoted definition of sustainable development: "development that meets the needs of the present without compromising the ability of future generations to meet their own needs."

Opponents of mining (and other industries) commonly claim that mining is not a sustainable activity. This claim might seem strange for an endeavour that has persisted since the end of the Stone Age, and it is easily countered:

• Virtually all the materials required by the technology on which our civilisation depends are the products of mining. Put simply, minerals left in the ground cannot support societal development, while biotechnology as yet is nowhere near providing substitutes – "what isn't grown must be mined."

DOI: 10.1201/9781003134008-2

- Restricting access to mineral resources is in no sense "development that meets the needs of the present" and might, in fact,[1] "compromise the ability of future generations to meet their own needs."
- There is no basis on which to conclude essential mineral resources will become limited in the foreseeable future, given continually improving exploration methods, our ability to mine deeper than before, continuous innovations in processing, and the ongoing expansion of recycling practices.
- When done responsibly, mining contributes to sustainable development, particularly its economic dimension. A mining operation typically provides fiscal benefits to the Government, supports economic growth through the purchase of goods and services and the payment of wages and benefits, and contributes to infrastructure development.

While widely used, the definition of sustainable development presented in the Brundtland Report is aspirational rather than precise, and the meaning of sustainable development and its usefulness as a concept have been the subjects of wide-ranging debate. A few years after the Brundtland Report was released, more than 300 definitions and interpretations of the concept of sustainable development had emerged (Dobson 1996).

This confusion notwithstanding, sustainable development has become widely accepted as the fundamental guiding principle for long-term global growth. The focus of sustainable development is the integration of environmental, social, and economic concerns into all aspects of decision-making, allowing business leaders to assess the sustainability of their companies and related activities. While some projects may not meet all sustainability objectives, environmental management standards and project closure have developed such that sustainability is achievable in most situations – companies indeed can *do well* by *doing good*. The key here is that we, as business leaders, should demonstrate that the benefits of our industrial activities outweigh the negative impacts.

---

1 Chapter borrows from "*Mining & Sustainability: The Three Circles of Sustainable Development*" by John Trudinger and Karlheinz Spitz, 2009; Title borrowed from "*Yes, Sustainability Can Be a Strategy*" by Ioannis Ioannou and George Serafeim, *Harvard Business Review*, 11 February 2019.

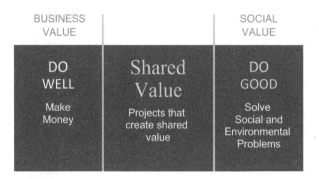

PROFIT, PEOPLE, AND PLANET

Three words summarise the traditional approach to business: Profit, Profit, and Profit. In the past, our engineers, and we as leaders, formulated, designed, and executed projects ensuring technical and regulatory compliance but paying little attention to social and environmental impacts and risks. Today's approach to business has changed; more and more business is about Profit, People, and Planet. Companies must consider their larger footprint. The smartest companies have learned how to do it in a profitable, environmentally and socially responsible way. However, profitability seldom happens by accident. Without proper planning and foresight, navigating environmental legislation and committing to social investments could drain the business dry, abruptly ending the transition toward sustainability.

In 1994, John Elkington, a famed British management consultant who would become a "sustainability guru," coined the phrase "triple bottom line" (Triple Bottom Line of Profit, People, and Planet, also known as 3Ps, TBL, or 3BL) as his way of measuring performance in corporate America. The idea was simply to manage the Company not only to earn financial profits but also to improve people's lives and the planet. A sustainable business works at the intersection of Profit, People, and Planet, and creates value for investors, customers, host communities, and the environment.

However, we face significant challenges in creating value for customers and investors without negatively affecting the environment. It can be challenging to switch gears between seemingly contradictory priorities, maximising financial returns while also doing

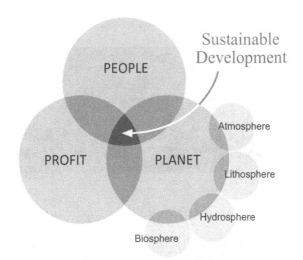

the highest good for society. Some of us might struggle to balance deploying money and other resources, including human capital, to all three bottom lines without favouring one at the expense of another. There is always a risk of unwittingly causing or worsening problems in one area while attempting to correct problems in another. In 1998, Robinson and Tinker postulated that the only sure way to avoid this is for business leaders to integrate decisions to consider effects in all three areas before taking action.

WHAT MAKES OUR BUSINESS SUSTAINABLE?

Economic sustainability is the concept of maximising the flow of income from a stock of capital while maintaining the stock yielding this income. The idea encompasses the traditional theory of economic growth; that is, receiving optimal returns from a given set of capital resources. The premise is that future generations will only be better off if they have more capital goods per individual than we have today. It is immediately apparent that population growth is inimical to sustainable development since it "dissipates" the capital stock. On the other hand, technological change enables a given capital stock to generate more well-being per unit of stock. An easy way to think of it is to say that future generations will be no worse off if capital stocks are "constant", while the rate of technological change can offset the rate of population growth. If technology progresses faster than population change, then future generations could still be as well off as we are today, or better off, with the same or even a lower capital stock. While this sounds airily theoretical, it is in fact an intensely down-to-earth and practical view of how the future can unfold.

The main message here and the core of Triple Bottom Line Theory is that capital goes well beyond the conventional idea of financial capital, and takes five primary forms (MMSD 2002):

•   Natural (or Environmental) Capital, which provides a continuous flow ("income") of ecosystem benefits, including biological diversity, mineral resources, forests, land and soil, wetlands, and clean air and water (primarily capital in the ecological system);
•   Built (or Productive) Capital, such as machinery, buildings, and infrastructure (roads, housing, health facilities, energy supply, water supply, waste management) (mainly all forms of capital in the economic system);
•   Human Capital, in the form of knowledge, skills, health, cultural endowment, and economic livelihood (small enterprise development, literacy, health care, inoculation programmes) (primarily capital in the social system);
•   Social Capital, the institutions and structures that allow individuals and groups to develop collaboratively (training, regional planning, decision sharing) (mostly collective forms of wealth in the social system); and
•   Financial Capital, the value of which merely represents the other forms of wealth (and the purest form of wealth in the economic system).

This broadening of the already broad concept of capital helps demonstrate how our businesses contribute to sustainable development – the rise in one form of wealth (e.g., human capital) offsets the decline in another form of wealth (e.g., mineral resources). In an ideal world, the increase of renewable capital through social investment would balance or outweigh the loss of non-renewable capital over the life of a project or business. For the extractive industry, the most critical challenge is to demonstrate that while we

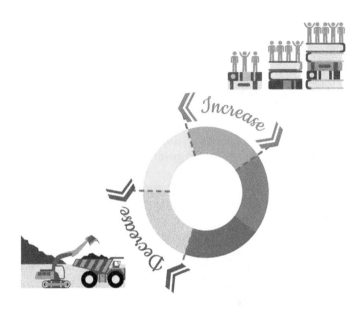

are exploiting stocks of natural resources, we are also creating sufficient renewable capital to balance the losses.

We face a second challenge, as we must aim to maintain or increase the level of capital stock and seek an equitable distribution of environmental costs and economic benefits. Social tension is unavoidable if resources and capital in the region of the project activity decline (depletion of mineral resources, environmental degradation, or disruption of societies) while capital stocks in areas outside the project area increase (tax income to the central Government, increases in shareholder wealth, or improvements in another region's standard of living).

Equitable distribution of renewable capital at the local, national, and international scales is challenging and is beyond the control of an individual Company. Throughout history, mineral resources have provided income for national Governments, but only a portion of created wealth and capital flows back to local communities. Companies are often blamed for the real or perceived unequal distribution of financial benefits at the local and federal levels, but much of the responsibility rests with the host Government.

International companies are also seen as exploiting national resources for the benefit of distant foreign recipients. This image is
grounded in reality, but the situation arises mainly in the context of
weak or venal national authorities.

Economic sustainability has clear linkages to the social system,
whether in addressing poverty through community development,
providing employment and business opportunities to host communities, or incorporating local input to economic decision-making.
Obvious linkages with the ecological system involve the supply of
raw materials for production (not only mineral resources, but also
air, water, and land) and the use of the environment as the final
waste sink (through emissions to the atmosphere and water, and
disposal of waste materials such as waste rock and oily sludge).
These linkages illustrate that economic sustainability boils down
to increasing one form of wealth while minimising the decline in
other types of wealth. As business leaders, we must refer to the familiar mitigation hierarchy: *Avoid – Minimise – Rectify – Compen-
sate/Offset* when designing measures to reduce the decline in any
form of capital (BP 05). The mitigation hierarchy neatly summarises common sense – where possible it is best to avoid negative
environmental and social impacts.

All resource development projects intrinsically involve
"trade-offs" between potentially conflicting goals, such as natural resources exploitation versus respecting traditional land
rights, or economic growth versus environmental conservation.
The challenge is to optimise trade-offs between and across the
three spheres fundamental to sustainable development – the ecological, economic, and social spheres. In summarising the significant challenges the mining industry faces, Sir Robert Wilson,
then Executive Chairman of Rio Tinto, urged the mining industry (in the June 2000 issue of *Mining Engineering*) to change its
dialogue with stakeholders, especially with non-governmental
organisations. He asked for the support of a new global mining
initiative to seek "independent analysis of issues that will determine the future of mining and that these issues are social, environmental, and economical." To depart from the traditional,
resource- intensive economic model, which implies a "grow first,
clean up later" mentality (Atkisson 2012), ecological considerations, and environmental protection need to stand at the core of
any business model and industrial organisation.

SUSTAINABILITY: COST THAT PAYS FOR ITSELF

Too many companies (and consumers) still see sustainability as a cost or obligation rather than a benefit, when it is, in reality, an opportunity. Today, investors regularly use Environmental, Social and Governance (ESG) metrics to analyse a Company's ethical impact and sustainability practices (BP 01 and 04). Investors consider such factors as a Company's carbon footprint, water and energy usage, community development efforts, and Board diversity. Data suggest that companies with high ESG ratings have lower costs for debt and equity, and that sustainability initiatives can help improve financial performance while fostering public support (Spiliakos 2018). "Doing good" can directly impact the Company's ability to "do well."

Is there evidence that the concept of "Profit, People, and Planet" has boosted shareholder returns or made companies one iota better run? Beyond helping curb global environmental challenges, climate change is one example of an opportunity for sustainability to drive business success (BP 13).

> If you really think that the environment is less important than the economy, try holding your breath while you count your money.
>
> – Guy McPherson

Here, energy saving is an obvious example and a "low hanging fruit." Energy represents a substantial portion of a business's costs, so reducing needless consumption will drive real savings. Experience indicates that a 30% reduction in energy consumption typically equates to a 10% cut in overall operating costs – attacking energy wastage isn't just right for the planet; the Company also benefits. Unlike many of the suggested solutions to combat climate change, technologies already exist at all levels to offer the insight required to save energy. To do this, however, energy consumption must be visible.

ROLE OF GOVERNMENT REGULATION AND LEADERSHIP IN INCREASING SUSTAINABILITY

Governments play a critical part in making sustainability integral to the private sector business, as much Government regulation is meant to provide protection, either to individuals or to the environmental commons. Whether the topic is environmental protection, occupational health and safety, or consumption of goods and services, Governments are in the best position to encourage or coerce sustainable and eco-friendly practices, and to promote sustainable behaviours. Policymakers encourage behavioural change and the uptake of technologies through incentives and legislation and regulation. As business leaders, we need to proactively monitor emerging Government initiatives to incentivise or force an industry's actions; surprises are generally costly.

There have been many attempts to codify the principles of sustainable development. With the possible exception of the Equator Principles (BP 04), the most widely known of these protocols is the United Nations Sustainable Development Goals. Within the mining industry, the International Council on Mining and Metals (ICMM) Mining Principles are commonly referenced. Although drafted for the mining industry, the ICMM Mining Principles would generally be applicable to all resource industries.

UN Sustainable Development Goals	ICMM Mining Principlesa
1. No poverty	1. Ethical business practices
2. Zero HUNGER	2. Sustainable development
3. Good health and well-being	3. Respect human rights and
4. Quality education	cultures, customs, and
5. Gender equality	values
6. Clean water and sanitation	4. Risk management
7. Affordable and clean energy	5. Improvement in health and
8. Decent work and economic	safety performance
growth	6. Improvement in
9. Industry, innovation and	environmental performance
infrastructure	issues
10. Reduced inequalities	7. Conservation of
11. Sustainable cities and	biodiversity
communities	8. Design, use, re-use,
12. Responsible consumption and	recycling, and disposal of
production	products containing metals
13. Climate action	and minerals
14. Life below water	9. Improvement in social
15. Life on land	performance
16. Peace, justice, and strong	10. Engage key stakeholders
institutions	on sustainable
17. Partnerships for the goals	development challenges

aAbbreviated versions; for complete ICMM Mining Principles, see International Council on Mining & Metals' website.

The UN asserts that mining has negative impacts on 11 out of the 17 Sustainable Development Goals. However, these goals and their associated targets are framed for Governments and are of limited relevance at the Company level. That said, they do provide a useful reference when drafting Company policies, goals, or long-term plans. ICMM states that the application of the Principles will support progress toward meeting the UN Sustainable Development Goals and the Paris Agreement on Climate Change, and a mapping of the 10 Principles against the Sustainable Development Goals is available. In 2003, ICMM partnered with the Global Reporting Initiative (GRI) to develop a Mining and Metals Sector Supplement to the GRI Sustainability Reporting Guidelines. This provides the basis on which ICMM members report their progress in meeting the 10 Sustainable Development Principles.

SUSTAINABILITY AND ITS CONNOTATIONS

The definition of sustainable development presented in the Brundtland Report is too wide-ranging to be applicable to a specific business case. Mining is a prime example.

At first glance, sustainable mining appears to be a contradiction in terms, as for all practical considerations, minerals are not renewable. Accordingly, all ore bodies are finite and thus, exhaustible. The once-proposed concept of "only mining the interest portion of the deposit every year" has no relationship to the real world of mine plans or setting of interest rates. Most mining projects have active lives of 5–50 years. However, the history of mining over several millennia indicates that new ore bodies are discovered as ore bodies deplete and projects close. Technical advances enable lower grades and different types of ores to be exploited. This leads to the virtually continuous development of new projects, so that supply of mineral commodities seldom lags far behind demand. Accordingly, the industry as a whole has proved resilient throughout human civilisation.

We should also be aware that sustainability has different connotations, depending on whose viewpoint is being considered. Some examples follow, again using mining for illustration.

< Mining Company – From a mining Company's viewpoint, sustainability means locating and developing mining projects to provide returns to shareholders and funding for exploration to

find or acquire replacement projects. Some companies, such as the Philippines' Benguet Corporation, once a significant producer of gold, copper, and other minerals, have sustained themselves during industry downturns through alternative revenue-generating activities such as real estate development. Another critical aspect of sustainability from the Company's perspective is to establish and maintain community support. Each project depends on community support, at least as much as the community depends on the project. It requires time and patience for a project proponent to establish trust, particularly in areas without a prior history of mining, and especially in areas with previous unfavourable experiences. It also requires consistency of approach, honouring commitments, and a willingness to seek out, listen to, and respond to all stakeholders' views. Ideally, the mining project becomes an integral part of the community, something of which the community is proud (BP 07).

As stated in the introductory section of this chapter, the mining industry itself can be considered sustainable, as there will always be ores mined. This follows because the elements that combine to form ores remain close to the Earth's surface, even after being used. When higher grade, readily accessible ores have been mined to exhaustion, lower grade and less accessible ores will be mined. And, in the future, particularly as production costs increase, more and more mineral and metal products will be produced by recycling.

< Governments – National, regional, and local authorities administer and regulate mining activities through laws and regulations that differ significantly from country to country. In some countries minerals are owned by the state; elsewhere minerals may be privately owned. Responsibility for regulation of mining activities may be at the national level, at the state or provincial level, or an even more local level. Commonly, different aspects of a mining operation are regulated at various Government levels. There may be overlaps, with more than one level of Government regulating the same aspect. Notwithstanding these differences, the Government viewpoint about the sustainability of a mining operation generally requires that the mining proponent must:

• Observe all applicable laws and regulations.
• Adhere strictly to terms and conditions associated with project approval, including environmental management and monitoring programmes.

- Implement and maintain community consultations and community involvement programmes.
- Avoid causing divisions within the local communities or adding to pre-existing divisions.
- Implement programmes to manage public risks associated with the project including delivery, usage, and storage of hazardous substances, operation of tailings storage facilities, and the use of explosives.
- Provide regular reports presenting up-to-date information about key project issues and the monitoring of environmental parameters.
- Ensure that sufficient funds are accrued or otherwise arranged so that the project site can be rehabilitated once mining has ceased.

Some jurisdictions, particularly for projects in remote or impoverished areas, also require proponents to assist the Government in providing health and education services and upgrading local infrastructure.

< Host Communities – The definition of sustainable development as "development that meets the needs of the present without compromising the ability of future generations to meet their own needs," hardly guides the aspirations of individuals or communities. From the viewpoint of a community hosting a mining operation, it is essential that the operation itself not be perceived as sustainable in the sense that it will continue forever. Again, all mining projects have finite lives, and communities should be well informed so that their expectations are realistic.

While communities exist that have been supported by mining operations for a century or more, in a great many others mining has ceased after much shorter periods. In some cases the associated communities declined substantially or disappeared totally, resulting in "ghost towns," the remnants of abandoned mining communities. This reality does not suggest there is anything intrinsically wrong with temporary communities. Many mines are developed in remote, unpopulated areas with no other potential sources of employment, so that there is no reason for the community to be sustained once mining ceases. This leads to "FIFO" (fly-in, fly-out) operations and Company towns built with relocatable infrastructure.

In many cases, communities that developed in association with mining have continued long after mining ceased, albeit on a reduced scale. Examples in Australia include many of the larger inland cities, such as Ballarat and Bendigo, which continued to exist and ultimately to thrive following conclusion of mining. What is vital to any community considering becoming host to a new mining project is that the community itself be sustained during and after mining. This usually means that the pre-existing livelihoods and economic resources bases are maintained, and that additional means of income generation are developed to replace mining once operations cease.

Different host communities have different requirements, expectations, and aspirations about new mining developments. Impoverished communities are likely to focus on employment opportunities, while communities that already enjoy high living standards will be most concerned with ensuring these standards are not eroded and their quality of life is not impaired.

< Employees – From the viewpoint of its workers, the over-riding requirements for a mining operation are security, meaning a reasonable expectation of continuity of employment for a defined period; adequate remuneration, commensurate with the circumstances; training as necessary to carry out job requirements and to provide opportunities for advancement; and a safe working environment. In some circumstances, employees will have additional requirements relating to family accommodation, health services, and religious observances.

< Shareholders – People and their institutions invest in mining companies primarily to make money. The expected return on investment is based on perceived risk, and the highest returns are demanded from projects with the highest risks. Risks traditionally have included sovereign risk, political risks, risks associated with delivering a project on-time and on-budget, and risks from geohazards such as earthquakes, floods, and landslides. To these must now be added social risks, which can range from chronic local opposition leading to schedule interruptions, to violent uprisings, or in the extreme case to civil war as occurred in Bougainville. Recently, there has been a trend toward "ethical investments" in which ESG issues are considered as well as the expectation of profits.

< The Environment – In a broad sense, "the environment" can be considered an essential stakeholder with potential to benefit from or be damaged by mining and associated activities. From the viewpoint of the environment, sustainability means that environmental values should not be lost or permanently degraded.

Environmental sustainability is a complex issue, involving much more than the rehabilitation of surface disturbance. Many involved aspects of the social and natural environment must be considered in evaluating sustainability: community livelihoods; indigenous cultures; community values; water resources, in both flows and quality; air quality; ecological functions; biodiversity; and sustainability of visual amenity. The measures used for sustaining or even enhancing these environmental components are the "building blocks" of environmental management, about which much information exists (Spitz and Trudinger 2019).

Much of the focus of environmental management aims to protect valuable environmental attributes and rehabilitate damage directly or indirectly resulting from development. Such rehabilitation usually seeks to re-establish the landscape and biota that were present before mining. While some mining operations can be totally rehabilitated, others leave part of the project "footprint," usually the "final void," that is not amenable to rehabilitation. Where significant environmental attributes are involved, permanent damage sustained in this residual footprint may be compensated for by environmental offsets. A typical example of an environmental offset in a forested ecosystem is the re-establishment of forest (by the mining Company) on land degraded by others, over an area exceeding that occupied by the final void, thereby providing a net benefit to the environment (BP 10).

Criteria for sustainability from the viewpoint of the environment could include:

• Conservation and re-establishment of representative vegetation communities and habitat types, particularly those associated with threatened or endangered species;
• No permanent net loss of valuable environmental attributes;
• Maintenance of hydrologic functions necessary for the maintenance of ecosystems; and
• Avoidance of pollution that could exceed the assimilative capacity of the receiving environment.

GREEN ACCOUNTING AND SUSTAINABILITY PERFORMANCE REPORTING

Environmental and social accounting is a direct consequence of the "triple bottom line" philosophy and forms the foundation of sustainability performance reporting. Environmental accounting, or green accounting, is the practice of incorporating principles of environmental management and conservation into reporting practices and cost/benefit analyses. Environmental accounting allows a business to see the impact of ecologically sustainable practices in everything from its supply chain to facility expansion. It enables business leaders to report on the economic impact of those decisions to stakeholders, to allow for proactive decision-making about processes that simultaneously meet environmental regulations while adding to the bottom line.

Why then, in an article published almost 25 years after Elkington coined the term "triple bottom line accounting," or Profit, People, Planet, does he want to "recall" the TBL framework (Elkington 2018)? Usually, recalls refer to faulty products to ensure public safety. Elkington did not recall TBL because he considers that focusing on economic, social, and environmental impacts is no longer needed. On the contrary, it is precisely because Profit, People, and Planet are so crucial that he suggests a recall. The main problem Elkington now sees is that the TBL has been reduced to an accounting and reporting tool that we cleverly use to show how great we are. As he explains, "Together with its subsequent variants, the TBL concept has been captured and diluted by accountants and reporting consultants."

Do we need to abandon his inspiring and straightforward theory? No. We can find ideas elsewhere about the true meaning of his pioneering thought. The theme of the 2015 Organisation for Economic Co-operation and Development Forum was "Investing in the future: People, Planet, Prosperity," with the term "prosperity" reflecting very closely what Elkington originally had in mind with economic impact. Economic impacts include the creation of employment, innovation, and paying taxes, all leading to prosperity. As business leaders, we can approach Elkington's original intentions with TBL: minimise our negative impacts and maximise our positive effects on the economy, society, and environment.

Business leaders demonstrate transparency as a sign of corporate responsibility. Efficient and reliable sustainability reporting is that which has been embedded into an overarching framework for Company performance reporting to shareholders. Sustainability reporting serves not only to control the quality of the Company's sustainability management but also to deal effectively with internal and external expectations of shareholders and the public at large. However, given societal sophistication concerning the environment, we must obtain and present material and reliable environmental and social data and information no matter the external pressures or internal limitations we face. Companies that can point convincingly to the integration of environmental management in their practices have a significant advantage in the public relations sphere.

Businesses often provide rich narratives on how they translate their vision for environmental and social performance into tangible business goals with measurable sustainability metrics. However, few can provide hard evidence that their business practices are not damaging the environment or the host community fabric.

The GRI claims to advance "the practice of sustainability reporting and enabling businesses, investors, policymakers, and civil society to use this information to engage in dialogue and make decisions that support sustainable development." To this end, GRI offers the GRI Standards, followed by over 70% of the world's largest firms, designed to create a common language for organisations to report on their sustainability impacts consistently and credibly. The standards are in the form of a modular set, starting with universal sustainability standards, followed by topic standards – economic, environmental, and social.

But what does "sustainable" even mean? To translate the Brundtland Commission definition into something meaningful at a Company or operational level is another matter altogether. And a lack of clear definition means there is also a lack of accountability.

Many of the GRI indicators are not necessarily straightforward. Take reporting on energy consumption and greenhouse gas (GHG) emissions. Increasingly companies switch to electricity from renewable sources, say by using solar panels, saving money, and emitting fewer carbon emissions. But are materials used in solar panels mined sustainably? Are panels recyclable? Does the solar energy Company effectively track its carbon emissions?

A Company may also report high local community engagement by having token programmes in all operations, but are these programmes effective? To demonstrate that community development programmes result in substantial positive community changes, companies must assess changes in the host community and a reference community to differentiate between general trends and programme success. Such broad and detailed data are seldom available. But the biggest problem with sustainability reports may be that they are often misaligned with Company priorities. Achieving success for TBL requires a careful balance between social, environmental, and economic performance. The measures for each, and how they are weighted, vary from Company to Company. Most companies face sustainability dilemmas where social, environmental, and economic needs compete. For example, developing a project, product, or service line may be beneficial for the society at large, but at a high cost to the Company in the short or medium term. Take discarded automobile batteries as an example. Batteries constitute the primary source of recycled lead. Superficially, considering that mining costs do not occur, producing lead from recycled batteries seems to be a very lucrative business. Reality differs. In many countries, there are still no incentives for battery owners to prevent uncontrolled dumping of batteries into the environment; supporting infrastructure to collect waste batteries is also lacking. Dismantling batteries is difficult, generating a wide range of undesirable hazardous wastes as by-products. Furthermore, lead recyclers often compete with primary lead produced by mining operations at low cost in countries with less stringent environmental laws and regulations. Only a few recycling companies will commit to lead recycling, although doing so would undoubtedly meet the Brundtland Commission's sustainable development definition.

How a Company manages its sustainability dilemmas or fails to do so may be far more illustrative of its commitment to the TBL than any sustainability report based on an externally imposed structure. Remember the history of financial reporting. Here, the publishing of annual reports predates any external requirement. In 1903, U.S. Steel published a yearly financial statement known as the earliest modern Corporate Annual Report. The guiding principle was to represent U.S. Steel's financial situation to its shareholders in the best way possible. The Company's story was not a report based on an external standard for annual financial reporting; these came much later.

From the outset, sustainability reporting has reversed the process, putting the cart before the horse. External standards and reporting formats limit companies' opportunity to tell their unique stories based on metrics and processes used to manage their distinct TBL dilemmas. Unless business leaders can align what is monitored and valued inside the Company with what is broadcast externally, sustainability reporting will likely remain little more than a public relations exercise, which puts rhetoric over action, or, at worst, snake oil sales talk.

SHAPING A MORE SUSTAINABLE FUTURE FOR OUR COMPANIES

For a Company or project to be sustainable means it meets the requirements and expectations of its significant stakeholders – shareholders, employees, customers, Governments, host communities, and the environment. The best outcomes occur when the needs and aspirations of these stakeholders align. Then, all are working for the same objectives and all share in the benefits. However, if one stakeholder group is too greedy and succeeds in obtaining a disproportionate share of advantages, then the sustainability of the Company and the entire project will be jeopardised.

The road to sustainability for most businesses is not easy. We business leaders can shape a more sustainable future for our companies and our communities by applying the suggestions listed below. It perhaps should be stated that this was written amid the COVID-19 pandemic, as we observed companies that prioritised short-term profits, and ignored planning for existential risks, disappearing (and not quietly) around us, or trying to survive on Government handouts.

< Assess What Sustainability Means for Our Stakeholders – To guide this process, ask questions, such as: How much waste is the Company producing? What impacts does our Company have on local communities? Can we reduce the energy bill? Do our hiring practices attract diverse job candidates and promote gender parity?
< Build Our Business on Belief – One can change everything about a Company other than core beliefs. Our Mission Statement should reflect the concept of sustainability; a clear Mission Statement is an integral part of a sustainable business. An effective Mission Statement outlines our Company's focus on

"doing" – a guiding light of why we do what we do. It needs to be consistent with what we, as a Company, are going to do to drive value. Our core beliefs are often reflected not in big but in small actions, such as finding ways to make our Company's meetings and business trips more productive and thus sustainable.

Choose real mugs, glasses, and plates over disposable ones

Combine several appointments on one business trip

Check your program for in-policy hotels with eco-labels

SUSTAINABLE BUSINESS TRAVEL

Use safe public transportation or carpool with your colleagues

If you rent a car choose an electric vehicle

(Based on Chefsache Business Travel)

< Change, Don't Stand Still – We should adjust the Company's strategy as necessary, always ensuring that the Company remains profitable. Profitability is always the number one priority, as no business is sustainable if it loses money. We cannot help our cause if we don't stay in business. But, as has repeatedly been demonstrated, practical sustainability efforts may help our companies to become more profitable. When adjusting strategy, we should focus on sustainability topics that are most relevant for the Company. When we change and act fast, our Company can be the boldest and brightest of its type; if we do not embrace the global shift to greater environmental awareness and sustainability, our Company may be brushing up against its risk of extinction.

< Manage Change – Establish a sustainability management framework to support implementing the sustainability strategy. Comparing management practices against other companies in our industry sector – benchmarking – will help us by illustrating the extent to which environmental, social, and economic sustainability is formally managed. And we should accept the hard reality that sustainable business and management comfort do not co-exist. There is no need to paint ourselves as a revolutionary business leader; being a cautious reformer will suffice.

< Focus on Shining in One Area – It is not about our Company delivering sustainability values to every stakeholder at every location all the time; it is about fitting into a broader sustainable business ecosystem. In the end, it is all about the sustainability contributions we are making to the larger economy of the firms that will survive well into the turbulent 21st century.

< Deliver – We must deliver, e.g., on resource efficiency, waste reduction, community development, gender parity, or site rehabilitation. We also must take a public stance on our results, not least through annual sustainability reporting as meeting a commitment to our stakeholders, not as image polishing. Impressive outcomes are unlikely to come immediately; the road to full sustainability is long, and business leaders may need to test different approaches to achieve the most significant impact.

None of the above will be simple, quick, or easy under any real-world scenario. None will be possible without investing resources – management time and attention, physical facilities, personnel, training, and outreach – that may produce few if any immediate returns. This will only happen if we start now and stay with it through to the achievement of desired outcomes.

How do ESG values change Company culture?

The notion that companies have cultures is rarely questioned, least of all by management consultants who generate much of their income helping organisations improve their culture.

Nevertheless, when employees or business leaders are asked to define culture, the responses reflect that culture, as it applies to a business organisation, is a complex construct.

Anthropologists – probably the only academics using the term in a reasoned sense – have never really managed to agree on precisely what culture means. E. B. Tylor, in 1869, defined culture as "that complex whole which includes knowledge, belief, art, morals, law, custom, and any other capabilities and habits acquired by man as a member of society," a definition still used widely today, but rather vague as applied to companies. The useful aspect of this definition, while not explicitly stated, is that culture seems to involve a group that shares something – *acquired...as a member.*

In business, the group is the Company. Undeniably, each Company has its own unique culture, one that is established consciously or unconsciously by the founders, which evolves in response to changes in circumstances, including leadership changes. Large multinational corporations tend to have similar cultures, reflected by similar attitudes and behaviours, driven by shared market expectations, and demonstrated by key employees. On the other hand, small companies, notably those privately owned, differ widely in their cultures, behaviours, and acceptance of risks.

What is the whole of the "knowledge, belief, morals, custom, and any other capabilities and habits" a Company may share, to come back to Tylor? The well-known saying that "A fish rots from the head" suggests that leadership is the root cause of an

DOI: 10.1201/9781003134008-3

organisation's failure, whether the organisation is a Government, a Company, or a business unit. How could it be otherwise? As business leaders, if we do not establish and protect a healthy culture, an unhealthy culture will likely fill the vacuum.

While most companies include environmental excellence among their core values, there are significant differences in the degree to which environmental performance is incorporated in the cultures and behaviours of different companies. Not all who "talk the talk" actually "walk the walk." To keep the fish from rotting, we must be smart and sufficiently concerned to critically scrutinise what the Company is doing – or not doing – and make the necessary changes, despite the pain that changes may create.

Within business in general and industrial companies in particular, there has been a slow evolution in environmental consciousness. In the early 1970s, when the requirement for environmental impact assessment first appeared, most companies considered it a nuisance – a hurdle to be overcome as quickly and as cheaply as possible. Subsequently, as environmental regulations became stricter and regulators more assertive, environmental issues assumed more importance among high-level managers and in the boardroom. Legislation in the USA certainly captured boardroom attention, as companies became aware of substantial environmental liabilities associated not only with existing operations but with old and abandoned projects and even waste disposal sites they never owned but used (further discussed in BP 12). Companies responded by recruiting more environmental staff and environmental lawyers, the latter bringing with them an adversarial stance toward Government regulators that remains to this day. More recently, management control has devolved to a younger generation of managers who have grown up with the environmental movement and generally do not question its legitimacy, though many believe, as a matter of political ideology, that regulations are always and everywhere excessive.

The emerging trend amongst progressive industrial companies is to embrace environmental excellence, not merely for compliance, but as a differentiator providing a competitive edge that potentially delivers benefits to both Company and shareholders. Such benefits may include reputation as a preferred employer, appreciation by customers, speedier approvals by regulators, preferential treatment when bidding for new properties or projects, inclusion in "green funds," and favourable responses from financial institutions.

Genuine environmentally and socially responsible attitudes as reflected in actions rather than words prove to be most effective in establishing and maintaining a reputation for Environmental, Social and Governance (ESG) excellence. When these attitudes and actions are perceived to be superficial, the "greenwashing" label may diminish the reputation.

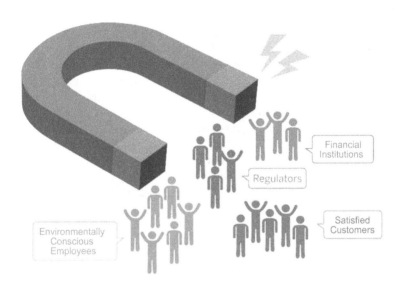

VALUES, PRINCIPLES, AND POLICIES

Most companies maintain a set of written statements outlining their values, principles, and policies with which all employees are expected to comply (Mission Statements, Codes of Conduct, Environmental Policies, to name a few). Large, multinational corporations may have formulated literally hundreds of policies and directives, and smaller companies generally much fewer. Policies and directives cover a wide range of activities and behaviours, including working hours, overtime, vacation, and sick leave entitlements; authorities for expenditure and reimbursement of expenses; sexual harassment; corruption; drug and alcohol; workplace safety; and many, many others. Many respond to Government regulations, industry standards, and insurance requirements as much or more than reflecting values, principles, and policies.

Today, it is relatively easy for even a small Company to create a website, sell goods and services worldwide, and adopt a Mission Statement with environmental and social corporate values as a top issue on the agenda. Read on to uncover why Company culture is much more than merely a Mission Statement, or a set of written statements.

Mission Statements

When Mission Statements were first introduced, they provided a useful and sometimes powerful encapsulation of what a Company wished to be. However, now there is so much similarity among these statements – they tend to tick all the same boxes, e.g., referencing environmental protection, as well as workforce health and safety, and community relations – that they no longer say much about the Company in question. This is not to suggest such statements have no benefit. For employees at any level, it might be useful for them to ask themselves, "Is this action consistent with our Mission Statement?" And for the public at large, the Company's actions may well be judged the same way. A negative answer to the question will cause reputational damage.

With the technology upstarts, a new generation of business owners has emerged, making significant changes and defining Company culture as what the Company, from an employee perspective, is like to work for. The Mission Statement of Google only reads: "Our mission is to organise the world's information and make it universally accessible and useful." These are not the companies where your grandparents worked, clocking out at 5 pm sharp, eyeballing the corner office. In fact, there is no more corner office – just hot desks and open floor plans, or working from home.

Companies like Google replaced the "rat race" with something quite different: co-workers are supposed to feel like family, hierarchies are flattened, and vacation days are unlimited (not that one would ever take them). And forget about work-life balance – today is all about work-life integration. Why else would companies provide for on-site massage, nap pods, and free dinner after 7 pm? And when entire professional staffs are for months working from home in a pandemic, perhaps integration is complete (and concepts of working hours and days totally evaporate).

Almost two-thirds of millennials "consider a company's social and environmental commitments when deciding where to work," according to an employee engagement survey by Cone (2016). Millennials, who comprise half of today's workforce, want to believe they are making a difference and look to their employers to support that view. For the millennials, a fancy environmental Mission Statement is quite insufficient. They want to see the types of behaviours they could mirror in their personal lives; as consumers, they exercise their options of making more sustainable choices. The bottom line for business leaders is to create a happy and fulfilling work environment where employees feel they can make a meaningful impact and want to stay with the Company. This can mean encouraging energy-saving and waste reduction and, as the COVID-19 pandemic has demonstrated, replicating physical meetings virtually, saving attendees time and reducing carbon emissions.

Codes of Conduct

Over the past 20 years, Codes of Conduct have evolved – from compilations of rules telling employees what they cannot do to guidelines for inspiring good performance. Here, conduct relates specifically to interactions between Company employees (and possibly employees of contractors and suppliers) and members of local communities. The aim is to encourage mutual respect and avoid behaviours or attitudes that offend local people or cause social disruption. Accordingly, a specific Code of Conduct is tailored to an operation's circumstances, and input must be sought from the relevant community or communities. Factors that might be included are "go–no go" areas for non-local employees, speed limits through local communities, fraternisation, alcohol consumption, social and sporting interactions, and observance of cultural and religious events. Some codes also include provisions to minimise environmental damage, such as recycling, wildlife protection, wildfire prevention, and protection of heritage values.

However, a Code of Conduct is only a foundation for behaviour. How it is understood, embodied, and practised defines a resilient corporate culture, something felt across the organisation, its business partners, and host communities. A Code of Conduct must be accompanied by a robust education and communication programme, as well as ongoing evaluation of compliance with the Code and its effectiveness in enhancing relations between the Company and community.

CULTURAL AWARENESS

When working across different cultures, it is vital to acknowledge the unwritten rules and cultural expectations of others. Failure to do so leads to misunderstandings. People may feel insulted that their credibility or stature is diminished, or, most commonly, that their viewpoints are not appreciated. Some misunderstandings may be innocuous; others will be serious. Think of culture as an iceberg – what is visible compared to what is hidden. The unseen facets of culture – values, beliefs, thought patterns, and perceptions – are the most difficult to understand and deal with.

Cultural awareness is essential wherever a Company operates outside its own culture, and most importantly, where Indigenous Peoples are involved (BP 07). Lack of cultural awareness and absence of cultural sensitivity on the part of staff working in unfamiliar cultures is a significant cause of misunderstandings, leading to disappointment, friction, and eventually conflict. Cultural awareness includes understanding traditional beliefs, values, customs, prohibitions, and taboos; religious beliefs and practices; family and broader group relationships and responsibilities; political systems, affiliations, and practices; and locally established conflict resolution mechanisms.

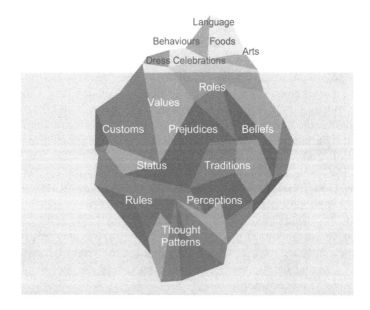

Perhaps the most important consideration here is to avoid the deployment of culturally insensitive personnel in situations where sensitivity is vital. This may prove difficult at times, notably during construction, as those characteristics that make for effective and efficient construction engineers may prove less than compatible with cultural sensitivity. The "Cultural Lens" is an ideal way to think about cultural differences – we all wear a pair of "invisible glasses" that shape how we see things around us. A lens functions to focus light passing through it and also serves as a filter. It limits the light coming through, thus minimising glare and confusing reflections, and enhances the perception of shades of colour. So perhaps the imagery of coloured lenses can be useful; indeed, the cliché "rose-coloured glasses" implies shading perceptions of reality.

Within the same culture, colours are closer together than where different countries or ethnicities are involved. Similarly, within the same industry, colours are closer together than where other sectors are involved. Very seldom will a construction manager wear Cultural Lenses that render colours identical to what a local community liaison officer sees. Here, the Code of Conduct becomes important, however unlikely it is to change the Cultural Lens of an uncompromising construction manager or anyone else.

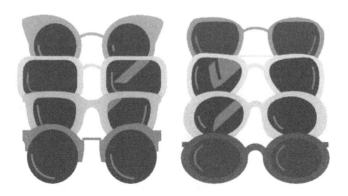

Cultural awareness training provides a valuable basis for interactions with local employees and members of local communities. Training in the local language is also highly beneficial, though requiring significant investments in time to be effective. Even better is when these programmes operate in both directions: local employees (and possibly even representatives of local communities) are provided with opportunities to learn about working arrangements, business practices, and socio-cultural observances of their expatriate visitors, and also to receive English language training.

Companies should articulate their own core values and codes of conduct, or failing that, endorse and implement codes produced by others.

World Business Council for Sustainable Development (1999)

GRIEVANCE MANAGEMENT

Despite the best intentions and efforts, there will be misunderstandings between an operating Company and the local community, and quite commonly these will lead to local people feeling aggrieved. Grievances not promptly addressed will grow and may lead to hostility and even conflict.

A Grievance Mechanism, a system of procedures for identifying, recording, and responding to community grievances, is now mandated as Principle 6 by International Financial Institution signatories to the Equator Principles (EP). The Equator Principles were established by the largest international banks to adopt International Finance Corporation (IFC)/World Bank Safeguard Policies as Applicable Environmental and Social Standards (Principle 3). In response, the IFC established its Performance Standards, as explained in BP 04. A necessary part of grievance management is that each grievance needs to be thoroughly evaluated, and the procedures changed if grievances are not being adequately addressed. Of course, this does not mean that all grievances will be resolved. However, those who feel aggrieved must see that their resentment is understood and that the Company responds, and the response is explained.

Applied effectively, Grievance Mechanisms offer efficient, timely, and low-cost conflict resolution. Used as an integral element of broader stakeholder and community engagement, a Grievance Mechanism can enhance local relationships and positively impact operational plans, schedules, and costs. Lacking a mechanism can render any and every grievance a potential crisis that could negatively impact all the same factors.

JOB DESCRIPTIONS

Senior leadership is not always readily observable, and employees will act as they are incentivised to do; this extends beyond short-term financial reward. Employees will quickly learn the "rules of the game" to survive and thrive at the Company and act accordingly, with actions that may have nothing to do with whatever aspirational values are posted on the walls and websites. Employees will practice the behaviours that they perceive to be valued in practice, not necessarily the values that are espoused.

As with health and safety, it is vital in any industrial operation that environmental performance is understood as the responsibility of all, not only the Environmental Department. This should be formalised in the Job Description for each staff member, and accordingly, environmental performance should be included in each individual's annual performance review.

ESG values constitute a core part of Job Descriptions for business leaders. To date, there may have only been a few cases of Chief Executive Officers losing their jobs following an environmental incident or controversy. Expect this to change with millennials' growing influence in business and evolving views on the accountability of business leaders.

USE OF EXPATRIATES

International companies operating outside their home countries most commonly employ expatriate staff in key positions. There is often a sound justification for this because large projects, especially during construction, demand a wide range of skills and experience that is seldom available in any one country. Usually, the General Manager is an "expat," and frequently, so are other managers. National professionals more commonly fill the Environmental Manager and Community Relations positions. These skills are usually readily available in most countries, and familiarity with local conditions and regulatory requirements are highly advantageous. Increasingly, there is pressure to minimise the number of expatriates, partly to meet local expectations and partly to minimise the costs involved in employing expats. As operations continue, local staff are trained to fill key positions, including management roles. Typically, after a few years, only two or three expats remain.

There is a risk at this stage that the expats are perceived as an "inner circle," and national managers feel they are not considered to be of equivalent status; salary differentials aggravate this situation. Hostility towards expats is common in some developing countries, and when it occurs within the management team can cause severe disruption to operations. As mentioned previously, cultural sensitivity is an important consideration when selecting expatriate staff; it is no less important in selecting local management.

YOUR ENVIRONMENTAL TEAM

In fashioning the corporate culture in relation to the environment, attitudes and leadership qualities of the CEO are paramount. Without total buy-in by the Board and upper management, environmental performance will be deficient, regardless of the calibre of the Environmental Manager and her team.

Clearly, the selection of Environmental Manager is also of utmost importance. In this position, communication and management skills are more important than technical ability. Highly advantageous is the ability to see through details, capture the essence of a situation, and communicate clearly and concisely – upward, downward, and outwardly. Candidates for the Environmental Manager position may include professional scientists and engineers with a wide variety of qualifications and experience. Clearly, so many disciplines are involved that no one person can be an expert in any but a small segment of the environmental spectrum. Even an environmental team of 10 or 12 will be unlikely to have all the expertise required to address all issues that may arise. Universities have responded to this situation by establishing "generalist" degrees in Environmental Science, wherein graduates receive broad training that is necessarily somewhat superficial in most environmental disciplines.

However, the adage "Jack of all trades – master of none" applies here. Specialist degrees and, even more, postgraduate qualifications are valuable in several respects: to promote rigour, engender self-confidence in the candidate, and enhance their credibility among peers, including fellow team members, Government regulators, academics, and Non-governmental Organisations. Interestingly, the speciality or discipline appears to be less important than having the qualifications. Many excellent Environmental Managers' specialities are in science or engineering fields unrelated to "environment" in the usual sense.

Having said all that, certain guidelines can assist in assembling an environmental team. Environmental scientists working for industrial companies are applied scientists. Training in one of the applied sciences such as agriculture, agronomy, forestry, or environmental engineering will be more relevant and provides a more "hands-on" team member than a pure science degree, even though in a recognisable discipline such as biology.

ORGANISATION AND REPORTING RELATIONSHIPS

There is a wide diversity of practices relating to where the environmental team fits into the organisation, the reporting relationships, and the decision-making authorities that apply. Companies with multiple operations will necessarily be organised differently from those with a single operation. In mining or petroleum companies at the exploration stage, environmental matters usually become part of the exploration team's responsibilities.

Virtually all mining and mineral processing, oil and gas, petrochemical, and power generation operations employ full-time environmental staff located at the site of operations. Large companies with multiple projects in development employ environmental staff to work full time on project studies. Smaller companies with intermittent study requirements usually depend on consultants for this function. Similarly, large companies may employ staff dedicated to project closure.

The size of the environmental team will depend on the size and complexity of the operations, and the extent to which environmental monitoring and reporting tasks are undertaken by inhouse personnel or outside consultants or contractors. In most cases, all environmental staff report to the Environmental Manager (or Environmental Superintendent), who, in turn, reports to the General Manager. Typically, the Environmental Manager is responsible for:

• Recruiting and, where appropriate, training or arranging training for environmental staff;

• Formulating, implementing, and monitoring environmental standards and policies;

- Input to employee induction and environmental awareness training for non-environmental staff;
- Routine inspection of operations to identify instances of non-compliance with environmental regulations;
- Designing and implementing an Environmental Management System;
- Monitoring atmospheric emissions and ambient air quality;
- Monitoring water quality, including wastewater discharge;
- Monitoring noise and vibration;
- In mining operations, rehabilitation activities including sourcing of plants (often operating a site nursery), cultivating, planting, and maintaining vegetation, and monitoring effectiveness of vegetation establishment;
- Organising and managing waste recycling activities;
- Investigating, trouble-shooting, recording, and reporting environmental incidents;
- Conducting internal environmental audits and cooperating with external auditors;
- Interfacing with Government environmental regulators; and
- Preparing internal and external environmental reports, including input to "Sustainability Reports" (BP 02).

Community Relations may also be under the overall responsibility of the Environmental Manager. Alternatively, the Community Relations Manager may report directly to the General Manager. In the past, many mining operations combined Environment and Occupational Health & Safety into a single department. However, with the increasing demands on both functions, this has become less common. For a large operation, Environment, Occupational Health and Safety, and Community Relations may have individual Superintendents, all reporting to an Environmental & Social Manager.

There are also operations where the Process or Metallurgical Department employs its own environmental specialists – for example, to undertake stack testing or other air quality monitoring. Similarly, in some mining operations, the rehabilitation team is assigned to the Mining Department, as this department carries out all the earthworks upon which the rehabilitation processes depend. The choice of organisational arrangements will depend partly on the nature of the operations and partly on the candidates' experience and capabilities for management positions.

Large companies with multiple operations usually have environmental and social staff in the Head Office. These may include an Environmental Manager and a Community Relations/Social Manager responsible for formulation of Company-wide policies, preparation of Sustainability Reports, and organisation of and participation in Internal Audits. Usually, site operations staff will have a "dotted line" reporting relationship with Head Office counterparts.

As well as the reporting relations between head office and operations personnel, the decision-making authorities at each level are important. The destruction of the Juukan Gorge heritage site in Western Australia (BP 01) has focused attention on decisions made by head office staff remote from the operations where these decisions apply. If local Community Relations personnel had authority to halt mining operations, a different and far less damaging outcome would probably have resulted. With more autonomy provided locally, head office personnel would give support rather than direction or control.

USE OF CONSULTANTS

As mentioned previously, the environmental and social team will not have all the capabilities required to address all environmental issues that will arise during the operating period and subsequent closure. Additionally, many jurisdictions require that certain tasks – notably environmental impact assessment and environmental auditing – be carried out by "independent" consultants. This draws on the large environmental consulting industry that includes local, national, and international companies. In developing countries, where environmental expertise is concentrated in universities, individual experts or teams are frequently engaged from these institutions. Additionally, Government-sponsored scientific organisations such as Australia's CSIRO can be retained to provide expertise in various specialities that may not be available from other sources.

EMBEDDING ESG VALUES IN THE COMPANY

Strong Company culture has been linked time and time again to organisational success. Look only at the last decade worth of data to see the impact culture can have on companies and the people that work within them. Then consider, how can ESG values be embedded in the culture?

Companies often provide a set of admirable values prominently displayed on walls and websites. However, these do not necessarily align with actual performance as revealed by observable behaviours. If ESG excellence is the objective, then consistent efforts must embed these values throughout the Company.

Firstly, the stated aspirational values should be revisited and updated if necessary. Next, a frank assessment should be made of the extent to which current behaviours and practices align with the stated values. An external coach could be useful at this stage. This assessment will inform the next stage, which will be to reinforce the positive behaviours and practices and change those that do not align with the values. This may be relatively simple, or it may require a major reorganisation, leadership change, recruitment of staff with different skill sets, and provision of additional funding.

If the objective is that ESG becomes a major differentiator, a Strengths, Weaknesses, Opportunities, Threats (SWOT) analysis will provide guidance. Internal factors that deserve attention include specific knowledge, skills, and abilities available within the Company, plus available resources of staff, financing, and technologies. External factors vary widely: characteristics of the industry sector in terms of main competitors, entry-barriers or trends likely to affect business; the market in terms of values, buying behaviour, including demographic information necessary to appropriately position and price products and services; social norms, values, and trends, including concerns about social and environmental perceptions in the minds of customers, shareholders, and investors; and the salient legal and regulatory forces affecting business operations. Evaluating corporate ESG can follow the principles of extending strengths and diminishing weaknesses, incorporating internal and external factors, which leads to four strategic options – expand, secure, catch-up, or minimise, as presented with examples on the opposite page.

Four suggestions for embedding ESG in the Company without straining business resources follow:

		External business environment	
		Opportunities	Threats
Internal business factors	Strengths	Expand e.g., use sustainability image to increase market share	Secure e.g., increase visibility of sustainability achievements
	Weaknesses	Catch-Up e.g., create sustainability flagship project	Minimise e.g., comply with minimum sustainability requirements

< Encourage Employee Involvement – Chances are the team has a few ideas for reducing, eliminating, or offsetting the Company's inefficiencies and waste. After all, employees are the ones immersed in day-to-day operations. Beyond soliciting feedback, consider creating Company-wide team challenges. When employees can take ownership of initiatives, they are more likely to be successful. This may not only save money but even identify new revenue streams, such as turning wastes into new products – possibly as opportunities for local businesses.

< Start with Small Changes – they can make a big difference. Track the Company's consumables over several days. Enlist a team to help. Take a close look at where the business is spending money on single-use plastics in both business lines and employee supplies. You may be surprised by what is found.

For instance, single-use plastic drinking water bottles quickly amount to tens of thousands of bottles per day in a mining operation. Many companies increasingly move to replace plasticware with multi-use water containers. These changes positively affect the environment, employee sentiment, and even a business's bottom line.

< Build Sustainability into the Culture – Sustainability initiatives are not a one-time thing. Keep the ball rolling and build it into the Company's culture. And make sure to share this message externally.

< Look Beyond One Company – Integrate environmental thinking into supply chain management, as partnering a collaborative approach to green practices can enhance sustainability performance (which is of course also an expectation of investors and lenders, as detailed in BP 04). Collaboration with neighbouring enterprises is another effective means of spreading environmental and social benefits.

ESG factors in project finance and M&A transactions

Our responsibility as business leaders is to measure and manage the financial performance of our companies. We do that to convert good investment decisions into great ones and to challenge those that could do better. Our choices, up to recently, naturally focused on avoiding or mitigating traditional business risks and impacts to successfully deliver profits.

Yet the world is changing, and we must change too, not because external advisors may say so but because our Company's future success hinges on it. Today, business leaders need to ensure that increasing global environmental awareness is incorporated into investment decisions.

The economics of sustainability (BP 02) and acceptance by local communities are essential to businesses that plan for growth. Today's young, environmentally savvy generation will tomorrow be our investors, customers, and clients if they aren't already. At the same time, raw materials are becoming more challenging to access, and penalties for even one noncompliance event can substantially reduce profits. In an extreme event, social outcry may put an end to our business.

> People are going to want and be able to find out about the citizenship of a brand, whether it is doing the right things socially, economically, and environmentally.
>
> Mike Clasper President of Business Development, Proctor and Gamble (Europe)

Historically, we have included environmental assessments in the due diligence and management of our investments. Past due diligence was limited mainly to assessing known, potential, and

DOI: 10.1201/9781003134008-4

contingent liabilities and obligations associated with the properties and operations of the target Company and its affiliates. The nature of the transaction and the laws and regulations to which the acquisition target was subject defined the scope and extent of environmental due diligence.

Today, measures and metrics exist that are wider in scope and embrace the ESG factors. As responsible investors, we must ensure our investment decisions continue to provide benefits to our shareholders, companies, and customers and protect the well-being of current and future generations. As detailed in BP 02, we cannot proceed with a philosophy of "Profit, Profit, and Profit"; these days, to retain our Social Licence to Operate, it must be "Profit, People, and Planet." Society is on a remarkable journey to advance real sustainability, and business leaders have a critical role in leading the sustainability agenda within business, especially with our investments.

Developing a clear understanding of the ESG credentials of counterparties or acquisition targets is not easy. However, as companies have to work hard to embed sustainability cultures and strategies, the impact of a newly acquired business that falls below these standards can have far-reaching ramifications. More in-depth analysis is required for ESG due diligence. This involves assessing the broader societal impacts of a transaction, e.g., a review of compliance with International Labour Organisation requirements, considering child and forced labour issues across the supply chain, or assessing reputational risk when acquiring a business that operates in a sector or country with a history of human rights abuses or bribery and corruption.

What other measures and metrics can we apply to incorporate ESG factors into our investment decisions? We can learn from genuine major scale investors, the International Financial Institutions (IFIs).

PROS AND CONS OF INTERNATIONAL PROJECT FINANCE

At an early date, IFIs recognised the need for appraisal instruments that take into account ESG issues.[1] The World Bank incorporated the environmental impact assessment (EIA) process into its public

---

1 This Chapter borrows from 'The Equator Principles III – What Matters and Why' by Karlheinz Spitz (2017)

sector investment decisions in 1989. Subsequently, the International Finance Corporation (IFC), the private sector arm of the World Bank, released an additional set of Environmental Safeguard Policies tailored explicitly to private sector financing. Multilateral financing institutions such as the Asian Development Bank (ADB) broadly adopted the World Bank's guidelines. They formulated policies and procedures quite similar in nature to the original World Bank Environmental Operational Procedure OP4.01. While subtle differences exist between these guidelines, they are the same book with a different title. The multilateral institutions tend to support their environmental assessments with state-of-the-art scientific and engineering analyses.

We cannot expect high-budget pragmatic approaches to project finance by IFIs that mirror those of IFC or ADB; to the contrary, experience has been that IFIs scrutinise investment decisions at great length if not necessarily in great depth. To illustrate, let us consider project finance of a multi-billion-dollar project in Indonesia led by a multilateral finance institution. Environmental and social performance was "securitised" to a level at which ecological impacts within a 50m mixing zone of saline water discharge into the ocean developed into a critical concern. At the same time, the IFI interpreted a wobbly showerhead at an emergency eyewash and safety shower station as a warning sign of a potentially flawed Occupational Health & Safety (OHS) system. Not all that surprising – the public judges IFIs by their investments more than other banks because they are created and exist to use public money to benefit business, people, and nature. Moreover, since we judge them against their internal measures and metrics, IFIs have a habit of trying to err on the side of caution.

Why then, do we seek project finance by IFIs? When working with business partners in low- and low-middle-income countries characterised by high political and social risks, international project finance helps to spread elevated project risks beyond the internal resources of our Company, joint venture, or consortium. It can also help retain influence over individual joint venture partners, such as national companies (a requirement in many countries when developing natural resource projects) and junior partners who might have less experience in controlling project risks. Furthermore, IFIs provide a "comfort factor" for commercial banks in providing finance. It is also a form of implicit risk guarantee (principally associated with approaching development finance institutions whose shareholders – e.g., Governments – may have national interests related to the investment). Last but not least, IFIs provide credibility for a potentially controversial project through association with internationally recognised environmental and social standards.

WORKING WITH EQUATOR PRINCIPLES FINANCIAL INSTITUTIONS

While the World Bank's approach to environmental assessment spearheaded environmental and social project appraisal until early in this century, it remained limited to projects in which the World Bank or IFC had a financial stake. This changed in June 2003, when four commercial banks, ABN Amro, Barclays, Citigroup, and West LB, drafted a set of voluntary guidelines to promote environmentally and socially responsible project financing, formerly called the "Greenwich Principles" because the banks met in this suburb of London.

The public comment period led to the name "Equator Principles" (EPs), hoping that the EPs would become the global standard for development, equally spread from North to South and East to West. The EPs met this expectation, as they now represent the standard framework for assessing and managing social and ecological risks in project finance. They are used (at this writing) by 105 signatory commercial banks and other financial institutions such as export credit agencies in 38 countries (known as Equator Principles Financial Institutions (EPFIs) or Equator Banks).

The original EPs, drafted with assistance from the IFC, very much reflected the IFC Environmental and Social Safeguard Policies to establish "applicable standards" (EP 3). IFC immediately began the process of upgrading these into a comprehensive set of eight Performance Standards for Environmental and Social Sustainability (IFC PS). After issuing the first set of PS, IFC then developed its Environmental, Health and Safety (EHS) Guidelines (general and industry-specific), by extensively updating the World Bank's 1990s *Pollution Prevention and Abatement Handbook*. Together with the IFC EHS Guidelines, the IFC PS generally considered the codification of Good International Industry Practices (GIIP) for environmental and social assessment and management.

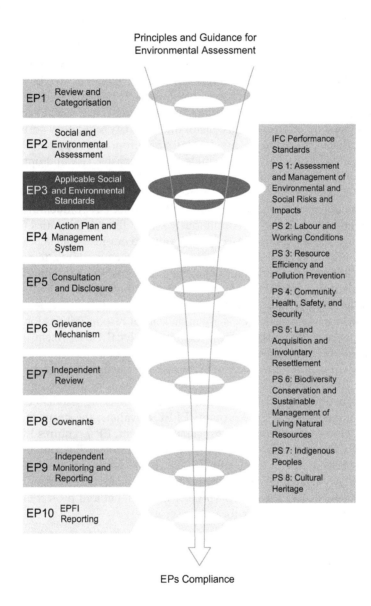

Principles and Guidance for
Environmental Assessment

EP1 Review and Categorisation

EP2 Social and Environmental Assessment

EP3 Applicable Social and Environmental Standards

EP4 Action Plan and Management System

EP5 Consultation and Disclosure

EP6 Grievance Mechanism

EP7 Independent Review

EP8 Covenants

EP9 Independent Monitoring and Reporting

EP10 EPFI Reporting

IFC Performance Standards

PS 1: Assessment and Management of Environmental and Social Risks and Impacts

PS 2: Labour and Working Conditions

PS 3: Resource Efficiency and Pollution Prevention

PS 4: Community Health, Safety, and Security

PS 5: Land Acquisition and Involuntary Resettlement

PS 6: Biodiversity Conservation and Sustainable Management of Living Natural Resources

PS 7: Indigenous Peoples

PS 8: Cultural Heritage

EPs Compliance

For most practical purposes, compliance with IFC PS and EHS
Guidelines equates to conformance with EPs and demonstrates the
adoption of GIIP.

IMPORTANCE OF RISK MANAGEMENT IN OUR BUSINESS

Performance Standard 1, Social and Environmental Assessment and Management System, reframes the way companies should handle social and ecological issues. No longer is it sufficient to conduct isolated environmental and social impact assessment (ESIA) studies, outsourced to external consultants to satisfy regulatory permitting/licensing requirements. Instead, if appropriately applied, PS 1 presents the standards as a single, comprehensive risk and opportunities management framework, fully integrated with the core of our business. The emphasis on risks and opportunities means that there is a need for us to not only avoid or reduce environmental and social risks but also to continuously search for opportunities that add ecological and socio-economic value to our investment.

PS 1 also emphasises our management accountability, to audit the adequacy of internal management systems and procedures to implement environmental and social mitigation measures outlined in the ESIA studies. Where found wanting, we may need to develop new business principles; clarify management responsibilities for engagement with workers, local community, local Government, and regulators; and implement procedures for long-term monitoring and reporting on the effectiveness of the risk management measures.

EPFIs also have responsibilities: EP 1 requires involved financial institutions to categorise the project considering associated environmental risks; EP 2 prompts EPFI to evaluate the status of the project's environmental and social assessment. EP 7 requires EPFI to conduct an independent project review (termed Environmental and Social Due Diligence or ESDD); EP 8 requires that environmentally related covenants be included in the loan agreement; EP 10 requires EPFI to publish project involvement and project environmental performance for public scrutiny.

PROJECT SCREENING AND CATEGORISATION

As part of their review of a project's expected environmental and social impacts, Equator Banks use IFC's system of project categorisation. The scrutiny applied to project financing depends on the project categorisation, with Category A Projects attracting the closest attention for due diligence review. Category A Projects are also subject to stringent inspections during construction and oper-ation. The question then arises, "Who determines the environmen-tal category of a project/investment?"

Equator Banks commonly rely on external expert advice com-bined with internal assessments based on a set of questions relating to the sensitivity and vulnerability of environmental resources in the project area, and the potential for the project to cause signifi-cant adverse environmental impacts. The size of the project alone does not determine the category, though substantial projects, and particularly those in extractive and other resource-based indus-tries, tend to be Category A.

The most environmentally or socially sensitive component of the project determines the project categorisation. When one part of the pro-ject has potential for significant adverse ecological or social impacts, then the project is to be classified as Category A regardless of the po-tential environmental and social effects of other aspects of the project.

As business leaders, we need to be aware that EPFIs generally tend to opt for a more stringent project categorisation regardless of actual project impacts. If project financing involves several banks, consortia tend to accommodate the viewpoints of each and Cate-gory A categorisation is often the only common denominator. EP-FIs also prefer to err on the safe side. If an information gap exists (e.g., about biodiversity), the worst case is assumed. EPFIs also play it safe as they are never criticised for applying a more stringent pro-ject categorisation. Finally, quite sadly, consultants appointed to carry out due diligence often tend to exaggerate social and ecologi-cal impacts to enhance their role, importance, and budget.

Several banks also define the types of projects that they do not finance, based on the World Bank and other multilateral financial institutions long-established "exclusion lists." Prominent examples include commercial logging operations in primary tropical forest, and production or trade in any product or activity deemed illegal under host country laws or regulations or international conventions and agreements, or subject to international bans. While Multilater-als' shunning of alcohol, tobacco, and weapons projects did not al-ways transfer to private financial institutions, some Equator Banks have distanced themselves from, say, financing deep sea tailings placement schemes or fisheries projects considered unsustainable.

Coal mining projects and coal-fired power plants, significant lines of business for many banks until recently, have also made their way into the exclusion lists of some banks. Coal use continues to decline in many countries due to policy, legal, regulatory, and market pressures reflecting climate change concerns and the increasing competitiveness of renewables. As business leaders, we are well-advised to anticipate how the global shift toward greater sustainability may affect our future access to project finance.

ROLE OF LENDERS' ENVIRONMENTAL AND SOCIAL CONSULTANTS IN ENVIRONMENTAL AND SOCIAL DUE DILIGENCE

Commercial banks and non-government-backed IFIs are generally the first port of call for us seeking international project finance. It is, therefore, vital to understand how these banks are interpreting the EPs. As stated above, many banks employ outside social and environmental specialists to explain how the IFC PS apply to their investment projects.

The people who developed the EP and IFC PS recognised the need for experienced practitioners. Unfortunately, not all consultants share this understanding. Some commissioning entities are undiscriminating when choosing advisors with the result that consultants with little if any relevant experience may be engaged. Workshops on adherence to the EPs are arranged based on a template format following in content and structure the IFC PS. Neither a site visit nor an understanding of the actual project is seen as necessary – a "cut-and-paste" approach from project to project substitutes for understanding project-specific and regional issues.

Does it matter? Yes. The environmental and social appraisal of our investments is not without challenges. A boilerplate approach to investment appraisal is implemented at the expense of all who have a direct interest in investment success, posing risks to the project developer, who likely will face project delays and will need to deal with ambiguity instead of clarity. It is also at the expense of involved bankers who seek informed, reliable advice, but receive "one-size-fits-all" guidance. Most importantly, it is at the cost of host communities that lose out on project opportunities that could have emerged had the EPs been applied in alignment with their original intent. It is also worth remembering that IFC PS and the embedded IFC EHS Guidelines are voluntary, not mandatory rules and standards. They were not meant to be prescriptive checklists, but are often applied that way.

As business leaders, we may, at times, need to challenge some of the unfavourable statements and conclusions made by consultants on specific issues. When necessary, this will typically require engaging the assistance of recognised specialised expertise in particular social or environmental disciplines. It helps to realise that IFI's retained consultants or in-house sustainability specialists may frequently be challenged beyond their depth of experience, particularly in countries where they lack first-hand knowledge.

The ESDD (EP 7 "Independent Review") is the evaluation carried out to assess compliance with a specific set of requirements and standards ("applicable standards" under EP 3). Lenders in project finance or companies considering a merger and acquisition (M&A) transaction conduct or commission an ESDD to ensure that a proposed investment does not carry unacceptably high social or ecological risks that could present a potential liability (including reputation risk) to investors in an M&A transaction, or lenders in project financing. Standards include our internal environment and safeguards compliance policies, and of course, the IFC PS and EHS Guidelines (including both General and Industry Sector EHS Guidelines).

EQUATOR PRINCIPLES AS EXAMPLE OF PRIVATE CODE OF CONDUCT

For a long time, public regulatory bodies issued all regulations. The traditional position was that Governments command the means and capability to supervise business activities and back them up with coercive power in compelling circumstances. Today, this form of governance is complemented by private Codes of Conduct and regulations introduced by business groups to self-regulate. Within the financial industry, the response to this governance evolution has taken the form of the EPs for project finance. As business leaders, we need to demonstrate to our lenders that our business decisions have found a balance in creating profit while protecting people and the environment. Compliance with EPs is not a legally enforceable requirement; it is a voluntary code for self-regulation. As is valid for all Codes of Conduct, EPs are process-oriented rather than outcome-oriented, and banks do not impose any penalties on their signatories' cases of noncompliance, except as provided for in loan covenants (EP 8).

COVENANTS IN FINANCIAL AGREEMENTS

While the EPs are voluntary and not legally binding, they take the force of law from the financing agreements into which they are integrated. In the case of noncompliance with EP requirements, the environmental covenants in the signed financial agreement are the basis for lenders to prescribe remedial measures and to invoke remedies.

The EPs since 2003 and the IFC PS since 2006 have changed from time to time, though the basic concepts have remained unchanged. Recent changes (EP IV in force since 1 October 2020) aim, for example, to reinforce the requirements for EPFIs to consider potential risks and impacts of projects, mainly in the areas of human rights, climate change (BP 13), and biodiversity (BP 10). The evolution of performance standards and interpretation within EPFIs can pose challenges for borrowers. Prospective borrowers need to keep pace with how EPFIs are interpreting the changes. It is usually uncertain which environmental and social performance issues might preoccupy the lending EPFI, starting with the financial closure (EP 4 "Action Plan and Management System" and EP 8 "covenants"), and then during monitoring and reporting (EP 9 and 10). It is essential for prospective borrowers to clearly define in financial agreements the accepted environmental and social measures and metrics to be applied at the time of financial closure. And while Equator Banks may not apply new requirements retroactively to existing projects, moving goal-posts are likely to apply to the financing of expansions and upgrades of such projects.

THE BIGGER PICTURE

The primary EP commitment is that EPFIs will not provide loans to projects where we, as business leaders, will not, or are unable to, comply with EP requirements. As such, EPFI will commonly screen the capability and the commitment of our business for implementing agreed environmental and social protection measures. A high-risk project combined with a low rating of the proponent, either in capability (reflected in a sound Environmental and Social Management System – ESMS, see BP 05) or commitment to implement agreed environmental actions (allocated resources and transparency), is likely to become a show-stopper for involved EPFIs.

We also need to demonstrate engagement with affected communities and stakeholders through disclosure of relevant project information, consultation, and informed active participation. Stakeholder engagement (EP 5) must match the level of project risks and impacts with particular attention paid to the involvement of vulnerable groups such as women, Indigenous Peoples and other ethnic minorities, low income and illiterate groups, youth, the elderly, and persons with disabilities.

We also need to realise that environmental and social aspects in project financing are sometimes "over-appreciated," for better or worse. Assessments and the EP application seem to follow a similar pattern, as was observed during the introduction of formal environmental management systems in the early 1990s. Initially, environmental management systems were seen not as management tools but as administrative systems. Companies, quite correctly, perceived environmental management system standards as too bureaucratic; too much bureaucracy blocked management systems from functioning correctly, as excessive paperwork "overkilled" the systems. Paperwork (though mostly digital now) is still a significant characteristic of accredited environmental management systems, but the initial overkill faded away as time passed.

EP compliance is unfortunately still often misunderstood as demonstrating adherence to each and every word in the IFC PS and related guidance documents – using them as checklists rather than standards and guidelines for performance. Current EP practice often seems not to differentiate between significant and insignificant risks and impacts. EPFI and their environmental advisors frequently opt for a "black and white" approach to environmental assessment and management. Project proponents need to demonstrate compliance

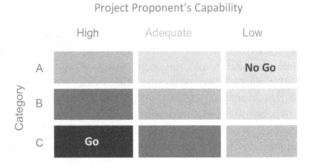

and adherence to every aspect covered in the EP framework, even risks and impacts that are in practical terms trivial or irrelevant.

As experienced business leaders, we know our sector and see our organisation's capacity to bring about change in a new project or acquisition. In project finance, we, and our lenders, should take a relaxed view on minor noncompliance issues that can be addressed in the initial phase of project development. Similarly, in cases of mergers and acquisitions, we can tackle minor noncompliance issues during the initial period of new ownership. Significant unknowns, however, or concerns with real potential to affect reputation or cause future disruption to operations, we must view more seriously. Above all, experienced professionals must be able to distinguish the former from the latter. In practice, we will experience various shades of grey when dealing with noncompliance with EP requirements. Unfortunately, evidence of overly precautionary assessments of environmental and social risks is plentiful; some examples follow.

> Indigenous Peoples – In projects in developing countries, it is often convenient to argue that host communities fall under the category "Indigenous Peoples." As a result, more stringent safeguards apply. In many countries, determining a rigid identification formula for Indigenous Peoples presents difficulties. Indonesia, for example, has a population of approximately 270 million. The Government recognises 1,128 ethnic groups; the 15 largest groups account for almost 85% of Indonesian citizens. The Government of Indonesia holds that the concept of Indigenous Peoples is not applicable, as practically all

Indonesians (with the notable exception of ethnic Chinese) are indigenous and thus all are entitled to the same rights. "Isolated, vulnerable peoples" is the term used officially by the Government to describe vulnerable groups that have virtually identical characteristics to "Indigenous Peoples" as defined in IFC PS 7. In some projects, raising ethnic identity through differential treatment based on ethnic heritage would do more harm than good.

Free, Prior, and Informed Consent – It is often convenient to argue that a project has not obtained free, prior, and informed consent (FPIC), as was incorporated into the 2012 version of the IFC PS. FPIC is a specific right initially acknowledged in the case of Indigenous Peoples, as recognised in the United Nations Declaration on the Rights of Indigenous Peoples (UNDRIP). There is no universally accepted definition of FPIC. It does not require unanimity and may be achieved even when some individuals or groups within or among affected Indigenous Peoples explicitly disagree. For project compliance with IFC PS 7, consent refers to the collective support of affected Indigenous Peoples communities for the project activities that affect them. It may exist even if specific individuals or groups are objecting to such project activities.

Protection and Conservation of Biodiversity – Since the revision of the IFC PS in 2012, there is an increasing focus on considering biodiversity impacts in the environmental assessment of new developments (BP 10). It has become fashionable to categorically argue that a project will impact biodiversity and ecosystem services, effects that are often exaggerated based on questionable assumptions. Any significant changes in conditions, whether natural or anthropogenic, can reduce biodiversity according to one or more evaluation standards. As biodiversity values are difficult to quantify, it may be simpler and more comfortable to argue that a project will impact biodiversity than to prove that it does not.

KEY TAKEAWAYS

ESG factors are increasingly important in project finance and M&A alike; the IFC Performance Standards allow for a systematic approach to ESG due diligence. However, aligning the interests of customers, shareholders, and regulators is difficult in a market where speed is often key to a successful bid.

ESG factors affect the likelihood of securing project finance. Poor performance of our projects in ESG terms may prevent project finance or affect lenders' willingness to provide lending. On the other hand, excellent performance on ESG factors will increase lenders' motivation to arrange project finance. Furthermore, poor performance in ESG factors will negatively impact the lender's assessment of risk and be used as leverage in negotiating higher risk premiums. The cost and difficulty of upgrading a project so that it meets EPFI standards can be significant considerations in closing project finance. The ease of integrating EPFI's expectations into the financial agreement is essential.

On the other hand, a thorough ESDD may lead to material cost savings; for example, more energy and water-efficient designs can significantly lower life-cycle costs. Our investments can also be made more resilient or "future-proofed" against emerging sustainability risks such as climate change or water scarcity. Incorporating sustainability issues into capital investment appraisal can aim to future-proof long-lived capital assets against these trends. Most measures for mitigating emerging sustainability risks are of marginal cost compared with the potential material costs they help us to avoid.

On a final note, whatever we are investing in, improving environmental and social credentials demonstrates to our stakeholders (including shareholders, lenders, regulators, customers, employees, and the local community) our commitment to responsible business. This will bring indirect financial benefits through improved stakeholder relationships and faster planning consents and other regulatory approvals. Experience shows that companies with higher sustainability performance benefit from lower costs of capital due to a lower risk profile and increased resilience.

Environmental and social planning and management

What's measured improves[1]

Global environmental awareness, initially emerging in the late 1960s, has picked up momentum in the past decade. In the USA, Japan, and Europe, evidence of sincere support for environmental protection is an old story. As economic development advances throughout the world, it is accompanied by increasing worldwide environmental awareness.

We also see that environmental awareness and distrust go hand in hand. Distrust of industry and extractive industry sectors in particular, by the general public and Government authorities alike, has led on occasion to environmental paranoia. Thus, requirements for environmental and social management are likely to become even more challenging, as risks are emphasised or even exaggerated, and standards become ever more stringent.

Ultimately, environmental and social management is about respecting human rights. Everyone has the right to a clean and safe environment and to have the environment protected for the benefit of future generations through both legislative measures and measures that we implement voluntarily (see BP 03 on Sustainability). Environmental rights may not be explicitly stated in constitutions or national law. Nevertheless, the right to life, health, safety, and freedom cannot be realised without a healthy environment.

Global environmental awareness has become an essential leadership quality. As business leaders, we understand that the need for environmental stewardship is driven by a wide range of considerations, including regulatory compliance, establishing

1 Quote by Peter F. Drucker.

DOI: 10.1201/9781003134008-5

and maintaining community support, meeting the requirements of lenders, and enhancing a favourable corporate reputation. But above all, by avoiding costly mishaps, sound environmental and social management leads to improved business profitability.

One simple and perhaps most all-inclusive definition of management comes from BusinessDictionary.com, which says that "management is the organisation and coordination of the activities of a business to achieve defined objectives." In today's business world, these objectives will mirror our economic, ecological, and social responsibilities.

To achieve our objectives, we design and implement an Environmental and Social Management System (ESMS) appropriate to the nature, scale, and vision of our business. Most businesses today claim to have an ESMS in place; all can point to having an Environmental Manager or Plant Manager explicitly responsible for environmental reporting, permits, compliance, and the like. They likely have the personnel and perhaps even a dedicated department to manage social and ecological business aspects. Some will highlight a formal waste minimisation programme or donations to host communities; others will cite a strong commitment to the environment and society through a carefully crafted policy statement. Even if your organisation has all these components, this does not necessarily add up to a working ESMS that effectively manages environmental and social aspects, minimising or preventing negative impacts (using the mitigation hierarchy) and amplifying positive effects.

Mitigation hierarchies have been used for over a century in natural resource management (for one example see BP 11 on Biodiversity). They include prioritised steps that lead to the best outcomes for people and planet: "Avoid, Minimise, Rectify, Compensate/Offset," adapted for the system to which they are applied. These hierarchies are inspired by Muir's Preservation theory (avoid/protect) and Pinchot's Conservation theory (minimise/compensate) – the basis of environmentalism in the USA. While usually applied to negative impacts, an equivalent mitigation hierarchy to amplify positive effects could read: "Create, Maximise, Foster, Compensate."

Managing a business's environmental and social aspects with corporate resources under a programme within an Environmental Department differs from operating an ESMS as a systematic approach to linking and managing all aspects of a business's operations and product interactions with the external environment and society. To this end, a Company that has implemented many elements of an ESMS may, for the most part, still be reacting to environmental and social issues rather than proactively adapting its ESMS in the face of rapid change.

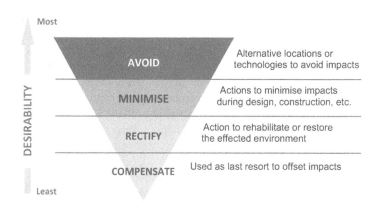

KEY ELEMENTS OF PROACTIVE ENVIRONMENTAL AND SOCIAL MANAGEMENT

The five essential elements of a proactive ESMS are social and environmental policy formulation, planning, implementation, evaluation, and review.

The Environmental and Social Policy articulates our commitment as business leaders to environmental compliance. It outlines the aims and principles for managing social and ecological issues. Businesses have increasingly decided to voluntarily formulate policies to demonstrate that environmental and social values are part of organisational decision-making. This also heightens awareness among employees, positively influencing behaviour and actions to achieve high levels of social and ecological protection.

Planning includes analysis of the environmental and social aspects of our operations. We identify ecological threats such as air and water pollution that can harm our host communities. We develop targets and programmes to reduce negative aspects (e.g., energy consumption) and to accentuate positive ones (e.g., social investment). We also delegate responsibilities, identify budgets and schedules, and present a broad picture as to how we will achieve defined objectives.

Planning, as a fundamental function of management, includes pre-emptive risk planning. One example is climate change adaptation, anticipating the requirements of a changing climate (BP 13). Climate risk is now widely recognised as an investment risk (Fink 2020). The goal is to reduce business vulnerability to harmful effects of actual or expected future climate (e.g., drought and extreme weather events). Risk management should be visible across the Company, and in this digital age, we can easily connect our Company with a single enterprise work management solution (BP 11).

Implementation converts our plan into action. We identify requisite equipment, resources, and employee training programmes to achieve the Company's environmental and social policies. A functional ESMS requires the participation of all employees, from top management to shop-floor workers. Importantly, ESMS actions and emergency preparedness and response planning and training need to be well documented and readily accessible to all.

Evaluation involves monitoring environmental and social activities and outcomes to confirm whether we meet our targets, identify any nonconformance, and implement corrective action(s).

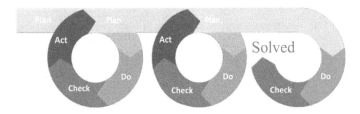

Plan what you are doing > Do what you said you would do > Check that you did it right > **Continuous**

Act on anything that went wrong to avoid errors of the same nature in future > **Improvement**

The ESMS reserves a special role for us business leaders, the regular review of the ESMS to adjust its various elements based on experience. This latter aspect introduces the concept of continual improvement and completes the cyclical process of Plan-Do-Check-Act (PDCA) (Tague 2005).

Not all ESMS are alike; goals and objectives will vary based on the location and needs of various stakeholders, unique to each business. Although the wording may differ, the four steps of PDCA are essentially the same for all ESMS. The essential elements of any management system are formulated in accordance with the international standard for ESMS – ISO 14001.

ENVIRONMENTAL AND SOCIAL BUSINESS ASPECTS

Business interactions with the environment are identified by reviewing activities, products, and services of our Company and assessing the potential for each of them to have environmental or social impacts, whether positive or negative. This leads to the establishment of objectives to increase positive and reduce adverse effects.

Certain environmental aspects are more significant than others because they impinge upon legal or regulatory requirements, affecting our ability to carry out our business. We maintain regulatory compliance by defining and monitoring applicable legal requirements, and by planning the efforts and allocating the human resources and expenses associated with permitting, reporting, and monitoring. Ultimately, such vigilance avoids or at least reduces the frequency and severity of regulatory violations and resulting sanctions and costs.

As business leaders, we also need to focus on improving the efficiency of our operations, not only to optimise productivity but also to reduce consumption and minimise waste generation, thereby reducing the potential for environmental damage. Greater operational efficiency usually involves a renewal of equipment and facilities, and improved design of production processes to minimise inputs (energy, water, reagents, and other resources) and waste outputs (emissions, effluents and solid wastes). Higher-level administrative efficiency may reduce legal liabilities and shorten permitting procedures by promoting better relations with regulators and communities.

PROACTIVE MANAGEMENT MEANS PREVENTION

Companies have different reasons for investing in an ESMS, but the most compelling is that it helps create a proactive management atmosphere. Active means prevention – and what are we preventing?

Wastes, pollution, and inefficiencies – all have negative impacts on our bottom-line economic performance and our competitiveness in the market. Proactive management also means promoting opportunities – training, employment, business opportunities, scholarships, community development programmes, improvements to public infrastructure, site rehabilitation, to name a few. All of these generate support from regulators and host communities, contributing to business sustainability. Successful implementation of an ESMS requires an initial investment and ongoing support but ultimately reduces overall operating and rehabilitation costs associated with adverse environmental, economic, and social impacts.

> What a business produces, how it buys and sells, how it affects the environment, how it recruits, trains and develops its own people, how it invests in the community and respects the rights of people - all these add together to form the impact of that business on society.
>
> Business in the Community (2000)

In most interactions of our business with its environment, there are things we know and things we do not know – Knowns and Unknowns. U.S. Defense Secretary Donald Rumsfeld famously opined about "known knowns, known unknowns, and unknown unknowns." While the context of that quotation was very different, it provides a compelling summary of decision-making reality under uncertain conditions. In today's world of cut-throat business, we business leaders are compelled to make choices every day that are impacted by factors known and unknown – and it is most important to recognise that we don't know what we don't know.

The Known Knowns we handle via the plan – emissions to atmosphere and water, fuel and energy consumption, workforce, social investments, to give some examples. But what about those various Unknowns? How do we account for those in business? The management answer to Unknowns is threefold.

First, we allocate contingency reserves that are approved, documented, measured, and therefore managed. We draw from contingency reserves where and when we need to, and this provides feedback for further planning. The purpose of contingency reserves is not to completely cover all Unknowns, but rather to reduce them to a level acceptable to stakeholders.

Second, we maintain an Emergency Preparedness and Response Plan to provide for the unexpected – the Unknown Unknowns, the events we don't know that we don't know about. And if one looks throughout business history, it is the Unknown Unknowns that were always the difficult ones to manage. The impact of the novel coronavirus (source of the COVID-19 pandemic) felt by all businesses worldwide illustrates the limits of providing for the unexpected, the Unknown Unknowns. Business leaders were suddenly navigating a broad range of interrelated issues that spanned from keeping employees and customers safe, shoring up cash and liquidity in the absence of revenue, reorienting operations, and navigating complicated Government support programmes, unfortunately often without success. Everyone's external environment was changing in directions and toward conditions that no one could confidently predict.

Third, to be comfortable with all the various types of risk – known, suspected, and emerging suddenly from darkness – we must prioritise the development of skills and competencies necessary to be adaptable, agile, and responsive in the face of Unknown Unknowns. The challenge is to develop strategies that will enable us to respond to these, whatever form they take. Adaptability is

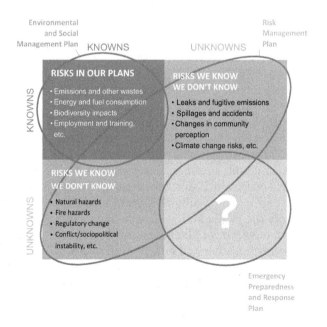

perhaps the most critical skill and mindset for dealing with wholly unanticipated contingencies. It is all about having and developing all necessary skills rapidly to adapt to a difficult situation and deliver the right response. We need the same process for any case that involves risk. Business leaders must adapt quickly to new challenges, rather than merely hoping for the best.

PLAN FOR THE UNEXPECTED

Change is often the only certainty – that is the unsettling reality of the 21st century world. Businesses face risks of which they could not possibly have prior knowledge. We business leaders attempt to address this problem by trying to identify as many risks as possible. While appropriate this does not address the fundamental "unknowability" of some challenges.

Despite all our best efforts, the possibility of unpredictable environmental accidents and emergencies will remain (BP 11, 14, and 15). Those that are predictable should be addressed in the ESMS through management plans and operational controls. Knowns we can plan for. The Emergency Preparedness and Response Plans provide for the unknown and unexpected.

A crucial component of the Emergency Preparedness and Response Plan is the Crisis Communications Plan (BP 14). Our business must respond promptly, accurately, and confidently during the hours and days that follow the onset of an emergency. We must be ready to address many different audiences with information specific to their interests and needs. If operations are disrupted, our customers will want to know how they will be impacted. We need to notify regulators; local Government officials will want to know what is going on within facilities they cannot normally access. Our employees and their families will be concerned and demand information. Community members living near our facility also need information – especially if they are threatened by the incident or are themselves in such difficulty that they could threaten the facility. Even before the media arrives, it will be demanding information, and media management planning must include social media.

Protocols for when and how to notify management should be clearly understood throughout our organisation and must be thoroughly documented and readily available. Consider an event that occurs on a holiday weekend or in the middle of a night. It should be clear to management and staff what situations require immediate notification regardless of the time of day – a business leader who must never be disturbed is nothing of the kind. Effective protocols and procedures must be in place to notify investors, regulators, customers, and other vital stakeholders. They and we never want to learn about a dire situation from news media. Public perceptions of how we handle the worst cases will positively or negatively shape our business image. If we do not communicate our circumstances, the void will be filled by others, usually by supposition and commonly inaccurate and damaging. This is not a time for finger-pointing or assigning blame; that will follow once the crisis is over.

DOCUMENTATION

While it is critical to initiate and execute all elements of the ESMS effectively, it is also essential to be able to demonstrate that we have done so. We must provide a steady stream of accurate information that enables those with a legitimate interest in our business to understand how we are managing environmental and social aspects. This information is essential for employees and external stakeholders – community members, regulators, customers, and other interested parties.

Documentation takes the form of records of various implementation and monitoring activities as well as ESMS results from training, audits, and management reviews. Everything needs to be reported alongside data showing the environmental implications of the Company's changes, e.g., waste reduction or optimising energy consumption. Records need to be managed to be easily accessed, retrieved, and presented meaningfully.

Sustainability reporting is our best effort to publicly document what we have done, and what we said we would do regarding environmental and social stewardship. Sustainability reporting is similar to other nonfinancial reporting forms: triple bottom-line reporting or corporate social responsibility (CSR) reporting. Sustainability reporting combines the analysis of financial and nonfinancial performance (BP 02).

ENVIRONMENTAL AND SOCIAL MANAGEMENT AS CONTINUOUS INTERLINKED EFFORT

Environmental and social management is a continuous effort throughout our operation's life – in the case of extractive industries, from exploration through planning and construction to operation and eventually closure. Ecological and social management activities in each part of the project cycle are linked to past and future operational phases. Poor understanding of our operations' environmental and social settings inevitably leads to deficient management plans, particularly when rapid, unexpected change arrives. And poor social and ecological management at any stage, but particularly during construction, may undo otherwise proper business planning and irreversibly taint subsequent phases.

NEED FOR ADAPTIVE ENVIRONMENTAL AND SOCIAL MANAGEMENT

Environmental and social management needs to be adaptive. Any dynamic natural and social environment is by definition never static, steady-state, or in balance. It will always change continuously in response to natural and anthropogenic forces, many of which we cannot accurately predict and of which we may be unaware. Similarly, despite extensive feasibility studies involving various sensitivity analyses and risk assessments, the progress of a project over time, and thus its impacts, cannot be reliably identified and quantified in advance. Add to this the unpredictability of community attitudes, political perceptions, and consequent changes in environmental regulation. It becomes clear that management plans must be adaptive and able to change quickly in response to changed circumstances.

Perhaps the most critical unknown in a profit-making business is the future price of the commodity, product, or service it provides. As commodity prices change, mine owners, for example, adapt by changing cut-off grades, production, grind size, reagent usage, employee numbers, and community development expenditures. All these changes have environmental and social implications. Lowering the cut-off grade may lead to changes in mine design which can result in more or less waste rock and more tailings produced over the life of the operations (Spitz and Trudinger 2019). It follows that project planning and mitigation measures need to be sufficiently flexible to cope with such changes during operations. This requires an in-built capacity to accommodate conditions beyond those predicted in the environmental assessment. Flexibility also means having contingency funds to finance emergencies or unforeseen changes in priorities or the external environment.

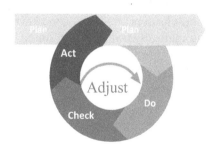

ENVIRONMENTAL AND SOCIAL MANAGEMENT AS BUREAUCRATIC APPROACH TO SUSTAINABILITY

Critics credibly argue that ESMS is a reflection of bureaucracy based on technical know-how and a deep awareness of the administrative workings of companies and Governments. Excessive bureaucratisation is undeniable when environmental and social management focuses on processes rather than on outcomes – a tick-the-box approach to managing environmental and social issues.

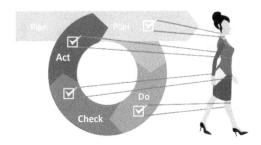

In contrast to a tick-the-box approach, the symbolic adoption of proforma environmental and social management, businesses with a substantive management approach will develop an effective response of reducing negative environmental impacts, by evaluating, managing, and controlling a wide range of effects with the primary aim of decreasing and even eliminating them. Such firms aim to be environmentally responsible, not just appearing so. What shareholders and stakeholders will not forgive are adverse outcomes arising because business leaders were focused on ticking a box. Good business leaders know that good governance is essential but does not automatically translate into sound business decisions.

Safety governance inside the boardroom

Board members are often geographically, physically, and mentally distant from day-to-day operations. And yet Board members' or senior business leaders' influence, tone, and culture set the emphasis on safety in any Company, through the questions they ask, their messages during interactions with employees, and their focus on organisational safety issues. In some companies, Board members interact predominantly with upper management and both groups have little or no contact with operational level employees. More enlightened Board members and senior executives actively seek to interact with employees at all levels, recognising that the most important insights are only likely to be gained at the "coal face."

We all readily acknowledge the upper leadership's essential role to foster a culture consistent with a safe, productive workplace in which human rights are valued and protected. Many Directors will be particularly concerned about compliance. Other Directors might ask about safety culture. How does our Company compare to its peers? How do others perceive our safety culture? Still others might focus on organisational sustainability questions, competition in the marketplace, and the impact of negative news coverage should a significant workplace incident occur. However, regardless of these considerations, most jurisdictions now place responsibility for worker health and safety in the hands of senior executives and the Board.

DIRECTORS AND PRISON

Generally, non-executive and even most executive Directors are not directly involved in day-to-day operations; they provide the

DOI: 10.1201/9781003134008-6

Company's overall strategic direction (BP 01). Safety is an exception. In many jurisdictions, Directors and executive managers have a direct legal duty to implement and monitor systems that ensure safe working conditions in their companies as far as reasonably practical. With statutory responsibilities to ensure companies do not breach Occupational Health & Safety (OHS) standards, Directors are every so often required to become directly involved with how their companies manage safety.

A fatality caused by a safety breach may result in the imprisonment of one or more Board members. One example was the conviction of former Massey Energy Chief Executive Officer Don Blankenship to a 1-year prison sentence related to a 2010 methane explosion killing 29 coal miners in West Virginia. In a first for Australia, in 2020 an auto recycling Company was convicted of industrial manslaughter following the death of a worker struck by a forklift operated by an unlicensed and unskilled driver. Two Directors were sentenced to 10 months imprisonment.

These convictions should ring alarm bells for all those holding Director-level positions and focus attention on the safety responsibilities that most regulators attach to those who govern businesses. Not all Board members fully appreciate the depth of their exposure to personal liability, nor that they may be held individually responsible for breaches of regulations, and can be prosecuted for severe criminal offences arising from their business conduct. At the Board and executive levels, it is not only about profitability, compliance, or statutory Company governance. For business leaders, work health and safety needs to command top priority. Good safety governance requires vigilance, ongoing OHS training and coaching, and proper workplace safety operating systems. Good governance practice ensures that every Board and senior management meeting has OHS as a topic on the agenda.

Some may think imprisonment is too harsh a sanction for business leaders who are not expected to engage in operational management. However, in China as one example, authorities take more stringent legal action. Yu Xuewei, the Ruihai Company Chairman, was sentenced to death after the Jiangsu Tianjiayi Chemical Plant explosion in 2019 killed 78 people and injured 617. His Vice Chairman, General Manager, and five other senior executives were sentenced to life imprisonment. The general rule in China's construction industry for many years was that three deaths in a single incident required jailing someone in management.

The top-down, command-and-control culture all too common within Chinese companies disconnects business leaders from the reality of safety on the front line. This also holds true for Boards in many other countries. The standard view of the Board's role, like the command-and-control management style, tends to keep executives in their offices, away from the action and unable to use their seniority to drive safety behaviours. Yet business leaders to some degree are personally responsible and accountable if workplace accidents happen. It does not need to be like this. It is possible to use the authority that the Board gives leaders to drive rapid safety improvements, if they learn how to direct it purposefully and skilfully.

SAFETY AND ENVIRONMENT

If we look back over the last 50 years, there is a discernible trend of safety and environmental incidents having an increasingly profound negative impact on business. Incidents that may have been measured in hundreds of millions of dollars in the 1970s would now be measured in the tens of billions. The BP Deepwater Horizon incident in 2010 cost BP over US$37 billion and took 36% off the value of the business. The Brumadinho Dam disaster in 2019 took 24% off the value of Vale stock, reducing the Company's market capitalisation by US$19 billion. Ultimately, severe safety incidents can have cost implications that even the largest companies can struggle to overcome.

Standing behind this phenomenon are several driving forces that increase societal expectations of companies regarding safety and the environment. The media and NGOs are increasingly savvy in identifying and broadcasting the consequences of incidents. Social media enable almost immediate communication of incidents in a way that would have been unimaginable just a few years ago. Gone are the days executives meet to strategise how they will communicate an unfolding incident. Now it is common that social media provide the first news of a safety issue over said executive's breakfast.

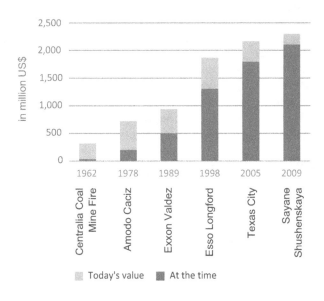

People are also better educated on safety and the environment than ever before, and air, water, and land contamination disasters have over time merged the two in the public consciousness. Pre-school children are learning about the environment. "Environmental projects" are now replacing "art and craft projects." Back in the 1980s, it was challenging to find a university course related to the environment; today, such courses are ubiquitous. Even lawyers and accountants have modules on safety and the environment as part of their studies. Millennials and the generations following are entering the workforce and the marketplace with a far greater awareness of these issues than ever before. They want to know the safety and environmental policies of companies that they join, and are intensely conscious of their personal safety and environmental quality. They also consider the environment in their investment and spending decisions (BP 16). Atop all this, "the difficult 2020/2021 pandemic years" have put everybody's public health and hygiene performance front and centre everywhere.

Improvements in science and technology provide another driving force. New and evolving technologies enable researchers to gather vast swathes of data on safety and environment, deepening the understanding of issues and the potential consequences of incidents at an unprecedented pace. Alarming images of massive explosions causing mass fatalities, victims of incidents searching for loved ones, collapsed buildings, and destroyed villages are within hours beamed worldwide, reinforcing the declining tolerance of risk within society and the associated response of Governments. As a result, safety and environment-related legislation is becoming more stringent, enforcement stricter, and mass protests commonplace. These forces continue to build. The media will keep reporting, social media are not going away, people will not forget their education, science will progress, and Governments will continue making laws. The expectations of society (and speed of switching to outrage mode) will continue to increase almost logarithmically. Companies are improving in terms of safety performance and response, but that improvement is not keeping pace with society's expectations. There is a gap between what companies are prepared to do and what society will accept. Those companies that can accelerate their improvements in safety and environmental performance and close that gap will thrive. Their next major incident may bring down those that do not. Given all this is true, what can companies do to promote and hasten improvements in safety and environmental performance?

HOW MATURE IS SAFETY GOVERNANCE AT YOUR COMPANY?

Ferguson (2015) argues that understanding the current safety governance level will help board members determine the necessary steps to move the Company to a stage where health and safety are fully integrated into business operations.

Levels of Safety Governance (based on Ferguson 2015)

At the lowest safety governance level (referred to as "transitional" by Ferguson), there is no clear understanding that "good safety" means "good business." Health and safety are seen as responsibilities of someone else, most likely the designated health and safety professional.

Compliance is the primary driver at the next safety governance level. A health and safety governance framework has been developed, characterised by basic (and often generic) safety policies and procedures. The Board is aware of its legal responsibilities, but not of the importance of their safety leadership.

After realising that compliance with legislation alone will not necessarily ensure everyone returns home safely every day, the Board will adopt a more focused approach to safety governance. A health and safety vision is introduced, and safety performance reporting will begin to include leading indicators. A health and safety management system is in place, focusing on the resourcing of the health and safety function and considering where the function is included on the organisational chart, so as be visible to the executive team.

At the next level, proactive safety governance often is driven by Board members who are confident in their safety leadership role. The Board may decide to establish a subcommittee to focus on health and safety.

The most effective level of safety governance occurs when health and safety are entirely integrated into operations. The Board acknowledges that a high level of health and safety performance is linked to business excellence. Throughout the organisation, safety committees share and obtain safety information from the Board subcommittee through to employee safety committees. The senior health and safety professionals understand their role is not merely technical but has a significant strategic focus for the business. There is a transparent sharing of safety data and learnings with other organisations in the industry and beyond.

POLICEMAN OR DOCTOR

The direct legal duty of Directors to implement systems that ensure safe working conditions is not the sole difference between safety governance and environment and social governance. Despite years of discussion and the wide availability of reporting frameworks (BP 2), the disclosure of material Environmental, Social and Governance

(ESG) issues remains a somewhat nebulous task, leaving companies grasping for ways to evaluate and compare ESG practices and risks. In contrast, workplace safety standards and practices have moved well beyond the currently disjointed potpourri of ESG assessment and reporting practices; Directors benefit from a sound body of knowledge on workplace safety and safety practices.

As with anything that we may care to measure, safety performance is a bell curve. Part of the organisation is excellent, part performs poorly, and the majority exists somewhere in between – compliance focused, in Ferguson terms. Unlike other things that we measure, safety depends on the weakest link, not the average. The majority of the Company may have a very robust safety performance, but it only takes one department, one team, or one individual at the bell curve's low end to initiate a significant incident. It is therefore critical to identify the areas where safety risks are highest.

Safety performance varies across teams, operating units, and divisions, with some areas being more fragile than others. These weak safety areas may be activity-specific; they could be endemic to regions or geographies or organisational entities and departments. If something goes severely wrong, it is most likely to do so at these points. Unfortunately, often these are the areas most resistant to change, most oblivious to risk and contemptuous of rules.

It is essential to appreciate that every team within a Company has its own culture and performance standards, the product of the actions, and especially the reactions of the team leader. The team's leadership quality is arguably the most critical factor in determining performance in safety and the environment, as well as everything else. Whether a team has incidents and accidents, whether it stays in compliance and follows the rules, and whether its members' behaviours are acceptable or problematic all depend on the quality of leadership.

Directors are tempted to point to their strict enforcement of rules, regulations, and directives to argue that they most certainly do not allow workers to behave in an unsafe manner. But accidents that occur tell a different story. A big part of the solution is to change the narrative around safety leadership. Companies that manage to shift leaders from enforcing safety as "police" to promoting safety as "doctors" can transform their safety culture. Peoples' instinct is to be wary of the police; tell them as little as possible, maybe even hide or cover up errors and mistakes, or point fingers and blame others, all of which makes for an unsafe workplace. On the other hand, we tell doctors everything – they look, listen, and diagnose and then work with us, prescribing the treatment and cure. The doctor's world is collaborative, empathetic, and healing – very different from the world of police.

This shift of narrative from "police" to "doctor" is a profound change of culture, requiring a new set of leadership qualities and skills. Leaders up and down the line need to learn how to engage and coach their people on safety in a way that creates connectivity, understanding, and learning. To understand what this means, they need to experience safety coaching for themselves, and this needs to start at the top.

Safety and the environment are more significant issues today than they have ever been and will be much more prominent in the future. They represent a source of increasing risk and opportunity for businesses that rise to the challenge. This can only be done through leadership, and leaders need to be better equipped for success.

SAFETY AND SYSTEMS

Much has been written on the benefits of formal safety management systems. The main benefits are demonstrating a good corporate image, building awareness of safety concerns among employees, implementing formal procedures to monitor regulatory and legislative changes affecting the Company, ensuring regulatory compliance, and minimising safety risks and liabilities. The figure below illustrates that a safety management system is essential to improve a Company's safety performance, but is not the sole solution. In fact, the existence of detailed systems and procedures can engender a false sense of security, and dilute the personal responsibility required to prevent human error.

The first step to safety excellence is responsible planning and engineering of production processes and activities. The second step toward success is having appropriate management systems in place during operation. Over the long term, however, safety excellence is based on a Company's culture and human behaviour. The point is that neither engineering nor systems can replace leadership and commitment. Buy-in from senior management and Board members is essential, which trickles down through the organisation, and eventually helps to develop a culture of care and safety and environmental responsibility.

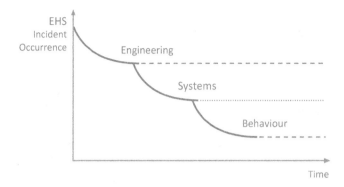

An EHS management system is a valuable tool to improve the EHS performance of a Company, but it is not the sole solution to EHS excellence

PREVENTING FATALITIES

In the context of good corporate governance, duty of care, and basic human morality, there is no outcome more important than the prevention of work-related fatalities. While work-related fatalities occur worldwide, fatal accidents in developing countries are more than five times higher than in developed countries.

A diverse range of opinions and approaches in theory and practice to fatality prevention is available. This text can only provide an overview of some of these. Minimising serious or fatal accidents is uniquely challenging. First, the causation pathways of fatal accidents and associated controls are often different from those of less severe incidents. Second, the frequency of fatal accidents within an organisation is typically very low compared with other types of incidents, leading to complacency or overconfidence within the management team. As a result, obtaining funding for programmes or studies explicitly addressing the risk of fatalities may be difficult. Third, there are usually few if any incident investigations involving fatalities through which controls can be identified and progressively improved over time. Finally, Lost Time Injury Frequency Rate (LTIFR) and Total Injury Frequency Rate (TIFR), ubiquitous safety performance metrics within the resource industries, are poor indicators of fatalities risk. It is not uncommon to find sites reporting good results for LTIFR data that have a history of fatal accidents.

THE PROBLEM WITH LTIFR

LTIFR is the most common metric of safety performance seen within the resources industries. Annual Reports and Sustainability Reports invariably include a chart of LTIFR performance over the years. Unfortunately, in both principle and practice, "lost time" is a poor measure of performance in managing the risk of fatal accidents, and not simply because supervisors frequently assign injured workers "light duty" to hide it where permitted.

The common problem with LTIFR as a measure of safety performance is the sample size. In well-managed and resourced operations, lost time injuries are relatively infrequent events. The resultant variability of LTIFR will likely make it useless for trending safety performance from year to year (put more simply, when accident rates are low, LTIFR may increase dramatically from 1 year to the next simply due to chance). This problem with using LTIFR as a safety metric of the risk of fatalities is fundamental. It cannot be addressed by statistical remedies such as aggregating incidence data for various sites or rolling averages (a technique that smooths out real change). Again, the mechanisms and chain of causation of fatal incidents differ from those of less severe accidents, as do the controls for minimising occurrence. LTIFR may at best measure the latter.

THE SAFETY PYRAMID FALLACY

First proposed by Heinrich in 1931 and updated several times since then, most notably by Bird in 1966, the so-called safety triangle has been used to support a model of incident causation in which serious incidents arise from conditions that allow lesser incidents to occur. Consequently, the elimination of less serious incidents will tend to reduce the occurrence of more serious incidents, up to and including fatalities. In this original conception, Heinrich theorised that there were 29 minor injuries and 300 non-injury incidents for every major injury or fatality.

More recent incident causation models have replaced this model. While the safety triangle or pyramid model is easily refuted, it is surprising how frequently it is reproduced in modern safety references. An organisation that is diligent in eliminating hazards and minor incidents is also likely to reduce more serious incidents.

Nevertheless, and to repeat, causation of incidents associated with fatalities is often very different from those of less severe accidents – the wearing of gloves to prevent hand lacerations will not prevent a fatality except under the most contrived scenarios. The safety pyramid is a dangerous model because it fails to identify the need for organisations to implement controls that specifically address the risk of fatalities and other serious accidents, increasingly referred to as "Critical Controls."

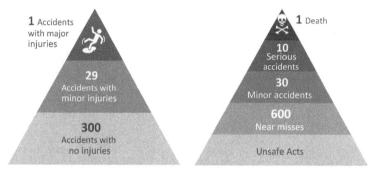

Heinrich's Pyramid 1931   Bird's Pyramid 1966

FATALITY PREVENTION MODELS

The following three models of incident causation and control are among the most useful for developing a systematic approach to minimising the risk of severe and fatal accidents. Because they address different aspects of risk mitigation, they can be integrated without issue.

Hierarchy of Controls – The hierarchy of controls is a systematic approach to mitigating workplace risk based on the sequential application of control methods, starting with the most effective and reliable method of control (top of the hierarchy) and, as necessary, moving step by step down the hierarchy to adopt less reliable methods of control. The hierarchy is based on the principle that actions to physically remove or isolate the hazard are inherently more reliable than controls dependent on worker behaviour.

The hierarchy of controls is a widely recognised methodology for risk mitigation referenced by various international standards such as OSHA, ISO 12100, and ISO 45001. Although the higher levels in the hierarchy may, in principle, offer the most reliable controls, in many cases, elimination and substitution are only feasible options during project planning. Well-designed engineering controls are highly effective in preventing severe accidents with little reliance on worker actions.

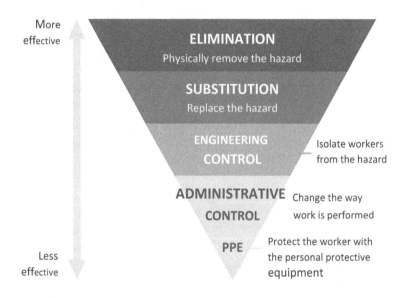

Layers of Control – Investigations of fatalities and other serious industrial incidents invariably confirm that there is never a single cause of an incident. A chain of causation is invariably involved in which a sequence of events is allowed to develop by the failure of controls at each step, culminating in the incident. The understanding is that, for a risk to express as an incident, multiple layers of controls must be breached. How these controls can combine in effect and provide redundancy when dealing with uncertainty is neatly expressed by the famous Swiss Cheese Model of incident causation.

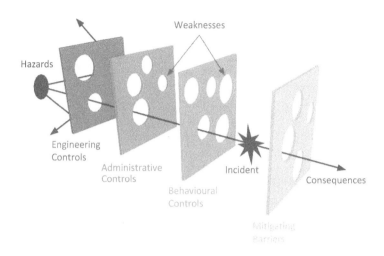

Interestingly, this useful and resilient accident causation model was developed by a psychologist rather than a safety expert (Reason 2016). James Reason's motivation in developing the Swiss Cheese Model was to integrate three common views of safety management: the person model, the engineering model, and the organisation model. In the Swiss Cheese Model, an organisation's defences against failure are modelled as a series of barriers, represented as slices of the cheese. The holes in the cheese slices represent individual weaknesses in individual parts of the system and are continually varying in size and position in all slices. The system as a whole produces failures when holes in all of the slices momentarily align, permitting "a trajectory of accident opportunity," so that a hazard passes through holes in all of the defences, leading to an incident.

Reason also proposed that while controls can exist at several levels within an organisation and take various forms, each serves one or more of the following functions:
- Create understanding and awareness of the hazards
- Give guidance on how to operate safely
- Provide alarms and warnings when danger is imminent
- Place barriers between the hazards and the potential losses
- Restore the system to a safe state after an event
- Contain and eliminate the hazards should they escape the barriers and controls

- Provide a means of escape and rescue should the defences fail catastrophically.

Critical Controls – The layers of control model does present some challenges. Many controls, such as risk assessments and engineering specifications, are not visible in the workplace and are poorly understood by most of the workforce. The identification and ongoing verification of the wide range of controls rest on implementing a complex management system. Even for a medium-sized operation, such a system will likely require dedicated staff and months to implement. There is also no clear basis for assessing the relative importance of different control types in preventing incidents, partly because the controls work and prevent the incidents. Data to support risk ranking of controls is likely very limited or unavailable when dealing with relatively rare events such as fatal accidents. And if all controls are included in such a system, it is by definition not risk-based.

At least in principle, these concerns are addressed by a model of incident causation and control based on the concept of Critical Controls. These are controls crucial for preventing significant incidents or "Material Unwanted Events" and minimising the consequences should one occur. In this model, the absence or failure of a Critical Control would significantly increase the risk of such incidents irrespective of the other controls' existence. The Critical Controls model is typically based on bow-tie analysis, a widely practised method for mapping incident causation and controls (this avoids the problem of limited incident history regarding issues relative to rare events) (see the Appendix).

The Critical Controls model is based on three general assumptions (ICMM 2015). First, the majority of Material Unwanted Events are known, as are the controls. Second, most serious events, including fatal accidents, are associated with failures to effectively implement known controls rather than not knowing the risks and controls. Third, the fewer number of controls, the more robustly they can be monitored.

Any approach that appears to offer a more straightforward means of addressing complex problems will always be popular. The Critical Controls model has received significant attention in recent years. As one example, International Council on Mining and Metals (ICMM) issued two guidance documents on Critical Control Management in 2015.

While the Critical Controls model offers significant benefits, like all other models of incident causation and control, it has its limitations. ICMM suggests that the "majority" of Material Unwanted Events within the mining and metals industry are well known, as are the "controls." For any mining Company committed to the elimination of fatal accidents, this would be an unacceptable premise. If the risk of fatalities is to be reduced "as low as reasonably practicable," then all credible fatality events must be known, along with required controls. Besides, the claim that most serious incidents are associated with failures to effectively implement known controls rather than a lack of understanding of the risks and required controls seems unprovable. If there was, in fact, a lack of understanding of the risks and required controls, how would this gap be knowable?

Safety isn't expensive, it's priceless.

Jerry Smith

It is also worth noting that under the Critical Control model, Material Unwanted Events occur due to the absence or failure of Critical Controls. However, this is not the same as saying that these absent or failed controls caused the incident. The Critical Control model's weakness is the attention given to the "intermediate" causes of incidents. The basic or root causes of an incident must be identified and addressed if the risk of incident reoccurrence is to be minimised. From this viewpoint, controls addressing these basic or root causes are the proper Critical Controls if the goal is to minimise the occurrence of fatalities. The model also essentially ignores the importance of elimination and substitution, the highest levels in the hierarchy of controls.

GOLDEN RULES

Workers, including new employees, are exposed to many safety hazards in day-to-day work at any large industrial operation. A Golden Rules Programme, common across the mining industry in particular, aims to eliminate basic unsafe work practices with the potential for leading to serious injury or death.

Although such programmes are relatively simple, several outcomes must be met to ensure success, as illustrated in the following using an open-cut mine in Indonesia as an example. The key objectives necessary to support the Golden Rules Programme were identified as follows:

- Ensure that all employees were aware of the major hazards in their work areas and required work practices that would, to a large degree, protect them from these hazards.
- Establish clearly understood sanctions for a knowing breach of these work practices, with consistent and fair delivery.
- Communicate this information in a highly visible and ongoing awareness programme.
- Monitor the occurrence of Golden Rule breaches and implement corrective actions as required to minimise the occurrence of such incidents.
- Ensure employee spouses, local community leaders, and local police and army leadership understood and supported the programme (bearing in mind that sanctions for breaches included final written warnings or immediate termination of employment, as appropriate).

The scope of the Golden Rules Programme was established following a review of similar programmes at other mining operations and a database of fatal accidents in the mining industry, together with an assessment of known major hazards at the site. Golden Rules were issued as pocket-size booklets, one rule per page, along with delivering an extensive awareness programme. A widely distributed Golden Rules Comic Book proved remarkably successful, with copies being spotted in surprising locations such as the local airport lounge and a karaoke bar.

From the beginning of the programme, it was made clear that the sanction for a knowing breach of a Golden Rule was a final written warning (with reduction of bonus) followed by termination of employment in the event of a second offence within 12 months (site contractors generally skipped the first of these). Concerning the applicability of sanctions (e.g., a driver failing to wear a seatbelt when operating a vehicle for a few seconds to adjust parking), a simple test was applied. Namely, did the breach place the employee or another employee at risk of injury?

In the first 6 months of implementation of the Golden Rules, there were nine recorded breaches. Two of these resulted in immediate termination of employment, and five resulted in final written warnings. Six of the violations were related to vehicle operation.

For employees coming from local rural communities in particular, with little or no prior experience of mining or industrial workplaces or the application of safe work procedures, Golden Rules provide an effective first line of protection against the most common Major Hazards at the workplace, with focus on communicating and embedding basic safe work practices rather than on implementing a sanction system.

Credit: PT Vale Indonesia

WORK STRESS AND CORPORATE RESPONSE

Some occupational health and safety risks have even entered the vocabulary in some countries as specific words, but still remain overlooked by many companies. *Karoshi* (過労死), which can be translated literally as "overwork death," is a Japanese term relating to sudden occupational mortality. Karoshi deaths may be heart attacks or strokes due to stress and a starvation diet, or suicide due to mental stress caused by overwork (People who commit work-related suicide are called *karōjisatsu* (過労自殺)).

Exhausted Japanese man sleeping on subway train, Kyoto, Japan. (Roberto Fumagalli/Alamy Stock Photo.)

Despite this, and although stress has been recognised as a significant occupational hazard by unions and Government OHS authorities worldwide, many companies pay little or no attention to this issue. Chronic misfits between employees' needs and capabilities and what the workplace offers and demands result in work stress that is not a disease or injury in itself, but can lead to mental and physical ill-health or in the worst case to death, and certainly contribute to workplace accidents.

Work factors that potentially cause chronic stress can be categorised as working conditions, job requirements, and work relationships. Working conditions are often easy to adjust to provide a better workplace – e.g., better equipment and work stations; increased security; improved lighting; or flexible work schedules, rest breaks, and predictable working hours.

However, the most stressful conditions at work often relate to management issues that can include lack of communication and consultation, increased workloads, marathon workdays, burnout, organisational change and restructuring, and job insecurity. The solutions to these stressors, which reduce to either dysfunctional work relationships or a mismatch between available resources and requirements to get the job done, ultimately rest with the Board, as both often reflect poor Company culture.

> This is the only place I know where they call a meeting at 5:30 pm on Friday to discuss work-life balance.
>
> Dr Harvey Van Veldhuizen, former employee of a multilateral development bank

The Board should promote efforts to find a better work-life balance for their employees. For example, Toyota now generally limits overtime to 360 hours a year (an average of 30 hours monthly). The COVID-19 pandemic has also made telecommuting more acceptable, de facto replacing commuting time with time for recreation or increased rest, and facilitating the care of children or other dependents. Other companies implement "no overtime days," or have introduced flexitime, allowing employees to fit working hours around their individual needs and to accommodate commitments outside of work. In Germany, meanwhile, major companies like BMW and Volkswagen have limited after-hours employee emails to combat a growing culture of hyper-connectivity. In the USA, leading investment banks like Credit Suisse and JP Morgan Chase have issued new guidelines to discourage analysts and associates (particularly the lower-ranking millennial workers) from coming to the office on weekends.

WHAT YOU CAN DO AS A BOARD MEMBER

A complex safety management system is necessary for the minimisation of risk in an industrial workplace. There are no short-cuts. Programmes addressing Critical Controls will add benefit to any safety management system, within the context of a broader range of controls that address all known causes or contributing factors associated with fatalities. Golden Rule, Critical Control, and comprehensive risk mitigation programmes such as Major Hazard Standards have a place in a safety management system given the goal is to eliminate fatal accidents.

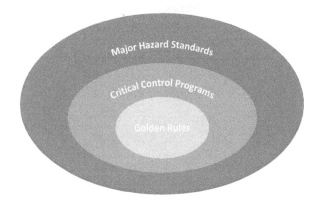

The Board has the most significant influence on addressing root causes, the failures from which all other failings grow, often linked to how an organisation is managed. The Board can reduce potential root causes of accidents by promoting and supporting initiatives to improve safety management within an organisation. Board members will rarely be on-site when an accident occurs, although they may in prior site visits observe or become aware of unsafe acts and conditions they could challenge. Hence, they must use their influence to establish a positive health and safety culture. And the Board must follow up on any serious accident or fatality.

Leading by example is likewise fundamental to embedding safe behaviours throughout a Company. If we, as business leaders, do not practise what we preach when it comes to health and safety, how can we expect employees to do so? However, talking the safety leadership talk is much less challenging than walking the walk. It demands the key change of no longer accepting things at face value. In essence, it's about walking the talk and creating a mindset change. It starts with building safety leadership capability at the Board and senior management levels, e.g., through safety coaching. It also demands that Board members and senior managers visit operations, reinforce the Company's safety system, and sell its safety mission. In addition, some simple steps will help to promote a safety-first culture proactively:

- Take the effort to understand Occupational Health and Safety, particularly the elements most relevant to your operations.
- Demonstrate personal commitment. Create a personal goal to engage in safety, such as leading an initiative to integrate safety into the Company culture.
- Conduct regular health and safety tours and take the time to understand the nature of operations (particularly where known safety risks are concerned).
- Ask challenging questions to alert all to the signals that indicate some conditions or practices need to change, more care is required in some areas, or where more resources must be employed.
- Demonstrate transparency through formal and informal communications. Celebrate safety successes and openly communicate safety challenges as they emerge.
- Test safety information received at the Board and check the follow-through on risk reduction activities (particularly those after an incident/accident).
- Ensure safety concerns are heard, and employees are included in safety planning, which may, in practice, require establishing a Board committee focused on safety.
- Establish an internal safety audit system for legal compliance.
- Have safety as a standing and prominent agenda item at Board meetings.
- Think strategically about safety, as it is not just about statistical analysis.

BP 07

Impacts of development on communities

Sharing benefits

We start this chapter with a key message: effective management of social impacts, positive or negative, requires knowing how to communicate with affected communities. It is simply impossible to engage meaningfully with community members without being a skilled communicator.

The key to becoming a skilled communicator is rarely taught in academia: enunciation, vocabulary, grammar, syntax, and the like are communication elements that focus on ourselves. When communicating with community members, more subtle aspects of communication that focus on others are essential. While successful business leaders might talk about their Company's plans and ideas, they do so in a way that also speaks to the host communities' emotions and aspirations. They realise that if their message does not take deep root with the audience, it will likely not be understood, much less championed. They focus on the leave-behinds, not the take-aways: by focusing on the community's wants and needs, business leaders learn more than they would by focusing on their own agenda.

DOI: 10.1201/9781003134008-7

SOCIAL IMPACT ASSESSMENT

The environmental agenda has evolved significantly over the five decades since the advent of environmental impact assessment. Initially, the main concerns related to pollution of air and water resulting from human activities in general, and industrial activities in particular. The effects of human activities on flora and fauna were also of concern. Social impacts were never ignored but received relatively little attention in those early years.

Attention on social impacts increased where developments were proposed in areas occupied by Indigenous Peoples. Unlike physical and chemical impacts, which usually can be effectively managed by applying well understood and accepted practices and procedures, social impacts proved much less easy to "manage." In fact, the idea of social management itself has unpleasant connotations and is a short step from social engineering. Although most people accept the necessity that the Government regulates some of their activities, the notion that constraints could be imposed on communities by an industrial corporation is not readily accepted. Many well-publicised cases arose where communities were adversely affected by large projects without receiving significant benefits. Livelihoods were damaged or destroyed while most employment benefits went to incoming workers, profits went to distant shareholders, and national Governments claimed taxation and royalty revenues.

Social impact assessment and the measures developed by corporations and Governments to address the impacts have evolved progressively to the stage that, for many developments, addressing social impacts is now of paramount importance. Today, social measures adopted by business leaders are many and varied. They include:

- Stakeholder identification, consultation, engagement, and participation programmes are designed and implemented from the earliest planning stages.
- Mechanisms are established by which grievances relating to project operations can be identified and actioned.
- Land acquisition is based on the principle of willing buyer/willing seller with fair and just compensation based on replacement cost, including due consideration given to economic displacement (BP 08).

- Workforce recruitment policies reflect community aspirations – usually by maximising local recruitment, together with appropriate training programmes.
- Policies for the purchase of goods and services favour local suppliers, including sponsorship and assistance for local parties to become suppliers and service providers.
- Codes of Conduct cover employees, contractors, and suppliers.
- Community development initiatives commonly involve water supply, sanitation, health and well-being, education, and capacity building for community groups, and fostering locally owned businesses.
- Assistance with and participation in community activities can include festivals, sports and recreation, and religious ceremonies.
- Preparation of employees, suppliers, and other stakeholders for the closure of operations starts with the initial planning, including finance guarantees.

It is vital that community members, including disadvantaged sectors of the community, are involved in formulating the policies and initiatives that will affect their lives. Those responsible for planning and implementing the project will have their own ideas, and some of the community's expectations are likely to be impractical, counterproductive, or unaffordable. Accordingly, negotiations on some points may be lengthy and complicated, involving trade-offs and compromises. However, handled well, there will be "win-win" outcomes with benefits shared between the project proponent, Governments, and communities. The project becomes part of the community, and the community becomes an integral part of the project. The most common threat to this ideal situation is greed. If Governments, communities, or corporations prosper at the expense of other stakeholders, disappointments will grow, and conflicts are likely to develop.

Government attitudes to industrial developments have also evolved in response to socioeconomic realities. Notably, most national Governments have (eventually) adopted policies whereby substantial portions of taxation and royalty receipts are used to benefit affected communities. Balancing national, regional, and local interests equitably has challenged many countries and will continue to do so.

As the above should make clear, experience has shown that a top-down approach to community engagement and development is rarely successful. We all tend to reject agendas set for us by others. Indeed, a top-down approach may adversely affect Community Relations by exacerbating economic inequities and social injustice.

Besides natural resources and capital goods, human resource enhancement is essential to achieve sustainable development that enlarges the range of choices community members can make about their lives. For reasons of both development and justice, deprived people, particularly women, youths, ethnic minorities, and the destitute, should have more power to shape their own lives. It follows that essential criteria for assessing social impacts on communities and designing measures to share benefits require an understanding of population composition (e.g., age, sex, and ethnicity), community concerns and aspirations, preparation for identification of grievances and resolution of conflicts should they occur, and the commitment to effective community development. The key to success is having the right people in place.

GENDER CONSIDERATIONS

On many issues, women and men may have quite different viewpoints and priorities. Therefore, it is essential that both genders be equally included in community consultations, engagement, and participation and that aspirations, interests, and concerns of both genders are similarly considered and reflected in development planning. This is not always straightforward. In some societies, men make decisions, while women are discouraged from interactions outside their immediate social group. Outsiders upsetting these practices are not appreciated by those invested in the status quo.

Another gender issue concerns the employment of women at all levels. Women leaders remain a minority, a statement that is no surprise to most of us. However, what is surprising is that men outnumber women in leadership roles across every sector globally: corporate, nonprofit, Government, education, medicine, military, and religion.

It is now not only acceptable to employ women in virtually any role within an organisation; it has also proved beneficial to do so.

Employers who favour the employment of men miss out on many benefits, including a wider human resource pool, improved business performance, happier workplace, reduced staff turnover, and better Community Relations. Institutional mindsets are the most significant barrier to the employment of women, as people make assumptions about women at work based on their stereotypical roles in society. Male business leaders should strive to unmask and correct hidden, reflexive preferences that profoundly shape our worldviews, affecting how welcoming and open a workplace is to diverse people and ideas.

INDIGENOUS PEOPLES

Public attitudes toward the extractive industries continue to change rapidly. For the mining and oil and gas industries, this accentuates the importance of establishing and monitoring good Community Relations and emphasising local communities' essential role as key stakeholders in any project. Nowhere is this more important than where Indigenous Peoples are involved.

Despite some different definitions of Indigenous Peoples, the following are widely recognised as indicative of indigenous status:

- Self-identification as members of a distinct indigenous cultural group and recognition of this identity by others;
- Collective attachment to geographically distinct habitats or ancestral territories in an area and to the natural resources in these habitats and territories;
- Customary cultural, economic, social, or political institutions that are separate from those of the dominant society and culture; and
- Distinct language or dialect, often different from the official language or dialect of the country or region.

It is evident that many if not most Indigenous Peoples have suffered through alienation, disease, loss of lands and livelihoods, and exploitation. While most non-indigenous populations in recent generations have experienced substantial improvement in basic human needs such as health, education, and welfare, such benefits have not always reached indigenous societies, particularly those remaining on ancestral lands. They remain among the world's most vulnerable people. While the rights of Indigenous Peoples have been widely recognised since the "Declaration of Rights of Indigenous Peoples" by the UN General Assembly in 2007, legal recognition and protection of these rights differ markedly in different countries.

If Indigenous Peoples are present in an area proposed for industrial development, and particularly if the development involves ancestral lands, Indigenous Peoples' issues will require sensitive and respectful consideration. While engagement with Indigenous Peoples involves much the same issues as engaging non-indigenous stakeholders, additional issues normally apply.

Many indigenous societies have been weakened by alienation from their lands, disease, exploitation, or misguided attempts toward integration. Once close-knit relations within communities and with their land and cultural beliefs and practices have been severely stressed as Indigenous Peoples find themselves in a cultural divide – not belonging fully to either the societal values of their own traditions or those of the larger society of which they have been made a part. Indigenous communities that remain may therefore be vulnerable to further social, cultural, and economic damage, particularly as traditional livelihoods become less viable, the physical culture is increasingly monetised, and generations develop contrasting or conflicting world views.

Legal and moral considerations aside, a compelling case for concern for Indigenous Peoples can be made on purely pragmatic grounds. Most, if not all, companies recognise the desirability (if not the absolute need) to develop and maintain good relations with the communities in which they operate. Before initiating contact with indigenous societies, it is advisable to seek assistance from appropriately experienced anthropologists.

In Australia, mining companies and Aboriginal communities often enter into consent-based land-use agreements which typically stipulate the conditions under which the Company can use the land and the compensation and benefits to be provided to the communities. In these agreements, traditional owners tend to be interested in risk reduction and compensation schemes, whereas companies tend to focus on obtaining tangible evidence of acceptance by their counter-parties of impending impacts. A range of payment schemes have been used, which all have their own strengths and weaknesses. Problems have arisen where financial support of communities is tied to progress toward or maintenance of the agreements.

For example, in October 2020, Fortescue Metals Group was accused of "routinely" failing to honour its agreements with an important traditional owner group and of withholding millions of dollars in royalties. It was reported that Fortescue had exploration permits over land subject to a land-use agreement with Wintawati Guruma Aboriginal Corporation (WGAC), which required there be no objection from WGAC when Fortescue sought to convert exploration leases into mining leases. However, WGAC had questioned plans to deal with culturally significant sites before the group would give its consent. Tribal elders claimed that Fortescue's response was to withhold substantial royalty payments on existing agreements (Australian Financial Review, 13 October 2020).

COMMUNITY CONCERNS

Issues of concern to communities faced with new industrial development will differ from one case to another, depending on the nature of the industry, the socioeconomic circumstances of the host community, and past experiences with industries – similar or otherwise. Despite the many differences, some concerns are common to most developments, including:

- Reduced water supply (BP 09);
- Potential contamination of water (BP 09);
- Land-use changes including the displacement of landholders (BP 08);
- Reduced air quality;
- Increased crime and disease due to incoming workers and/or people seeking work; and
- Lifestyle changes, including corruption of traditional norms, values, or morals due to incoming workers and increased economic activity.

Once construction and operations commence and the industry's nature becomes more apparent, initial concerns may turn out to be unfounded, while new concerns may emerge. Increased traffic, noise, or dust may emerge as the primary concerns, as would any unfulfilled expectations.

COMMUNITY ASPIRATIONS AND EXPECTATIONS

People will not welcome new industrial developments in their vicinity unless they perceive that benefits far outweigh any adverse impacts. Common aspirations that new developments will be expected to meet include:

- Direct employment of local people with opportunities to be trained in new skills;
- Increased business opportunities providing additional jobs; and
- Improvements in infrastructure, health services, and education.

Local Government authorities will look for a broadening of the economic base and increased revenues to support improved Government services.

Proponents seeking community approval should avoid any statements or suggestions that could lead to unrealistic expectations. Communities that have not had previous exposure to industrial developments may have highly inflated perceptions of the benefits that will occur. Such pre-existing expectations should be identified during community consultations and dampened down if present. Predictions of benefits should be conservative, so there is reasonable certainty they will be achieved. It is also most important that the information provided before project implementation be consistent. Communities are confused by mixed messages, and trust in the proponent may be reduced as a result. Even more important is that, as the project proceeds through construction into operation, all previous commitments are honoured. Promises should, therefore, be conservative so that they can be readily achieved or exceeded. Clear and accessible documentation of commitments is vital so that all those with a need to know are aware of all commitments that have been made.

A common situation facing proponents of industrial developments in relatively undeveloped areas is that most, if not all, local jobseekers are unskilled. Accordingly, local people are employed for cleaning, cooking, gardening, and security roles while the skilled, higher-paid employees are recruited elsewhere. This has been a source of conflict at many industrial operations and is mostly avoidable. Experience has shown that the lack of skilled candidates does not mean a lack of the necessary aptitudes. Training of employees in advance of project development has proved highly effective wherever it has been attempted. Benefits include more local employees meaning a more stable workforce, higher local incomes meaning more wealth in the community, lower costs of providing travel and accommodation for non-local workers, and lower risks of detrimental social effects associated with incoming workers. For companies with multiple operations, this may be readily achieved by on-the-job training at an existing facility. Where this is not available, most countries have a range of training institutions where most of the relevant skills can be learned.

CONFLICTS

Much of the preceding discussion is aimed at avoiding conflicts between industrial operations and host communities. However, despite the best efforts to minimise adverse social impacts and produce social benefits, conflicts can and do arise. Potential causes are numerous. Not all individuals in a community will benefit from the industry in their neighbourhood. Those who feel disadvantaged or left out may become disaffected, as may others whose expectations were not met. The Grievance Mechanism is designed to provide early warning of potential conflicts, but this process is not always successful at identifying grievances, and even when it does so, grievances will remain and grow if the Company is not perceived to have responded appropriately.

The Company's local employees will usually be the first to become aware of dissatisfaction with the Company's activities and often prove the best source of such information – a further reason to ensure locals are employed and not only at the lowest levels. An open communication system that encourages frank discussion at all levels within an organisation is valuable in this respect. Exit interviews may also identify grievances. Once a grievance is identified, it should be rapidly evaluated, and a response prepared. If the grievance is unfounded or cannot be fully rectified, the situation should be explained to those involved to understand the Company's position. Commonly, in these situations, an agreement can be reached.

If conflicts develop that cannot be resolved by discussions among the participants, external mediation or adjudication may be required. Informal mediation by community leaders may be appropriate. In addition, most societies have their own conflict resolution procedures, which are usually fair and perceived to be so by local people. In some societies, judicial appointees hear evidence on both sides and direct an outcome. In other societies, justice is dispensed by elected officials, while in indigenous societies, tribal elders may adjudicate. In any case, it is usually best if locally accepted conflict resolution procedures are followed.

COMMUNITY DEVELOPMENT

Most businesses, including industrial operations, provide support to local communities. In developed economies, this generally involves charitable donations, sponsorship of sporting teams, the arts, or other voluntary contributions. In developing countries, and mainly where new industries are being established in relatively undeveloped areas, community development is far more critical and is now seen as an integral part of Corporate Social Responsibility (CSR). In the past, community development programmes were devised by companies based on their own perceptions of community needs, often largely for the benefit of the Company and its employees. This was considered quite appropriate, considering that the initiatives were voluntary. However, more recently, such approaches are seen as patronising and self-serving. Furthermore, community development programmes are no longer seen as voluntary, and their scope and effectiveness receive much more attention. Community development efforts are now mandated in many countries, either by regulation or by conditions imposed as part of the environmental approval process. This introduces an additional bureaucratic element whereby Government approval is required for community development programmes.

It is now generally accepted that the host communities should participate in the formulation of community development programmes, and this is now the case for most recently established projects. Generally, the programme for each year is decided in advance to provide necessary input to the annual operational budget. Inevitably, differences will arise between the Company and participating stakeholders, each with their own interests and "pet projects." Sometimes these differences can lead to divisions that are difficult to reconcile. In such cases, an independent third-party, acceptable to the Company and stakeholders, can be commissioned to carry out a Needs Assessment to inform programme formulation.

The Needs Assessment involves extensive consultations with stakeholders to compile a long list of needs that are costed, ranked, and prioritised. Then the Company, in consultation with community representatives and the Needs Assessment consultant, selects from the highest-ranked needs to the extent that the budget allows.

Companies with community development experience have found that the best outcomes are achieved where the community itself contributes to the programmes. For example, the Company supplies materials and equipment, while the community supplies the labour. Ideally, community development programmes should emphasise training and capacity building rather than direct financial support. More specifically, programmes should be designed to strengthen the four crucial determinants of development shown in the following figure: income generation, education, and health improvement, infrastructure support, and capacity building. It takes years to build the quality and depth of community that will define societal resilience and health.

Most Community Development programs involve one of the following: income generation, health and sanitation, education and training, public infrastructure support, capacity building, and environmental programs.

Employment
Agribusiness
Cottage Industries
Entrepreneurship Training
Business Partnership
Development
Business Inoculation Programs

CAPACITY BUILDING
Decision -Making Culture
Long-Term Planning
Institutional Support
Civil and Economic
Education

INCOME
GENERATION

Family Financial
Planning
Well-Being
Programs
Emergency
Assistance

PRODUCTIVITY
INCREASE

EDUCATION AND
HEALTH IMPROVEMENT

INFRASTRUCTURE SUPPORT
Transportation
Sanitation
Access to Drinking Water
Village Meeting Facilities
Culture and Religion Support

Formal Education
Non-formal Education
Educational Institution
Development
Health and Hygiene
Awareness
Health Facilities
Development

SELECTING THE SOCIAL TEAM

Community Development is not exactly synonymous with Community Relations, and neither should be confused with Public Relations. Community Relations focus on fostering beneficial relationships between the Company and the communities in which it operates, while Public Relations are more concerned with generally promoting the Company's image and is accordingly more of a marketing function. This distinction is essential in selecting the most appropriate staff to represent the Company, both in Community Relations and in the community development initiatives as they are selected. The most effective people in these social roles are empathetic, with good listening and communication skills. However, they must also be pragmatic and realistic. Government Relations is another matter, but obviously all the "relational" staff and management need to communicate and coordinate with each other and keep senior management aware of relevant issues.

It is generally beneficial if local people are recruited for these social roles, as familiarity with local customs is essential. Besides, though it may not be obvious to all managers, local and particularly indigenous employees could be among the best candidates for environmental and social positions. It might even be feasible to select bright young people who have grown up in the local natural environment for polytechnic or university scholarships to prepare them to serve in such careers.

However, there is the risk that local employees, through family, business, or political connections, can have real or perceived conflicts of interest that affect their performance. (Similarly, such conflicts can arise in hiring local employees in security roles.) Some communities are seriously divided on racial, religious, or other grounds, and informed sensitivity is required to avoid the attendant biases, prejudices, and even potential conflicts.

In assessing the appropriateness and effectiveness of the project's Community Relations efforts, Directors might pose the following key questions:

During the planning stage:

- Have stakeholders been identified?
- What level of community consultation is proposed?
- Will stakeholders be invited to participate in project planning, and if so, on what aspects?

- Has opposition to the project been identified, and if so, how extensive is the opposition, and what reasons are stated by opponents?
- What commitments have been made?

During construction and operations:

- What is the ongoing community consultation programme, who are the participants, what is the schedule, and how are the results documented?
- What actions have been taken to identify and respond to grievances?
- hat grievances have been identified?
- What are the elements of the community development programme, how is the budget established, and how are developments selected and prioritised?
- Is there a process by which the success of the community development initiatives is evaluated?
- Have all project commitments been achieved? If not, why not?

Access to land as a human rights issue

Land acquisition in developed countries, while not without challenges, is relatively straightforward. Transaction of land title is between knowledgeable landowners and buyers, in an administrative environment that objectively documents land ownership. Some sellers will reinvest money from the land sale in other properties or long-term investments. Others will try to realise long-standing dreams, but without jeopardising their future livelihood.

Land acquisition in developing countries is seldom straightforward. No two countries are the same, and practices and procedures involved in land acquisition differ from one country to another and even within the same country. In some jurisdictions, Government authorities set the price for land sales, while in others, parties are free to negotiate. Formal land titles often do not exist; buyers and sellers must agree on traditional land ownership and rights (e.g., right to access or use). For many traditional landowners, land is not a mere commodity, but a source of livelihood and typically the only source. Private and shared rights in land, pasture, forest, water, structures, stocks of domestic animals, and food are essential physical assets. For these traditional landowners, the land is central to economic rights, often linked to their identities, and therefore tied to social and cultural rights.

In some societies, such as Melanesian communities in Papua New Guinea, inheritance laws have resulted in there being multiple owners for even small parcels of land. Identifying and locating all owners, while necessary, may be highly difficult and time-consuming, complicated by complex systems of kinship, absent relatives, and established traditions beginning to be questioned.

DOI: 10.1201/9781003134008-8

114 Environmental Social Governance

Residents whose land has been sold must relocate. Imagine you are being resettled. What thoughts might you have at that moment? You don't want to move because you love your home and neighbourhood? You don't have the money to buy a new property? You don't know where to keep your animals and grow your crops? These thoughts make the point that resettlement is not restricted to the physical relocation of people, but also includes the loss of physical and non-physical assets, including productive land and income-earning assets and the loss of cultural sites, social structures, networks and ties, and cultural identity. The term *solatium*, Latin for comfort, solace in English, is used in international and some national guidelines for land acquisition. It refers specifically to compensation for emotional loss and distress; some regulations assign it a monetary value.

Maybe most importantly for many traditional indigenous land-owners, wealth is more often perceived in non-monetary terms, such as family relationships and attachment to land and its natural resources. Indigenous Peoples are generally much less materialistic and have a different appreciation of money compared to western societies. Saving money obtained from land sales and investing in large or long-term productive ventures is difficult in those cultures in which individual wealth accumulation is not highly understood or valued. Further, the extended family or clan may have an automatic claim on any wealth received by an individual.

LAND AND HUMAN RIGHTS

Groups protesting large-scale land acquisitions regularly invoke international human rights law, reminding companies of their human rights obligations toward traditional landowners. Opponents understand that, in project development, the risk of violating human rights is never greater than during land acquisition. Rights to land include many levels of land tenure and use, all of which are relevant in the determination of physical or economic displacement: ownership, lease-hold, rights of use (such as for cultivation, grazing, and gathering), customary or usufruct rights (which include grazing, hunting, and fishing rights), ancestral domain (which refers to lands occupied or used by Indigenous Peoples), and third-party proprietary interests such as sharecropping, rental, or other formal or informal agreements. Systems of shifting or rotational cultivation (swidden), in which the farmer retains some rights to harvest residual tree or perennial crops, add to the complexity. "Land ownership" in the western sense is only one aspect of rights to land.

Source: Actionaid 2015.

RESETTLEMENT PRINCIPLES

Today the goal of resettlement is crystal clear: that displaced people will be better off after they have moved (IFC 2002). Since resettlement seeks to provide restitution by improving the livelihoods of affected people, a business should look on resettlement as an opportunity rather than a threat – opportunity to achieve community development, win the local communities' support, and earn the social licence to operate (BP 07). Ideally, affected people will share project benefits over the project's entire life span and beyond.

Therefore, there should be no confusion over the goal of resettlement: it is a valuable opportunity to reduce poverty. At least at the material level, "better off" means providing the very standard compensation package such as better houses, improved water supply, equivalent agricultural plots, access to educational and health facilities, and access to affordable energy. People in remote regions of developing countries are often impoverished; improving their well-being is rarely expensive and is well within the means of project proponents. If not, the project should not proceed.

Ideally, land acquired should be compensated by replacement land of the same or better size and utility. This is not always achievable, as purchasing replacement land may lead to additional displacement. Also, it is common that sellers prefer to receive cash rather than replacement land.

MINIMISING NEED FOR RESETTLEMENT

We know that avoidance is better than compensation and mitigation. The priority should be to minimise the need for resettlement, generally through project design, such as intelligent alignment of access roads or siting of supporting infrastructure. The location of some project components is fixed (an ore body is one example), but everything else is subject to project design. The project designers should investigate all feasible alternative options if large numbers of families are affected. Resettlement should proceed only if it is still considered unavoidable after all alternatives have been exhausted. An adequate environmental assessment needs to document the alternatives considered.

ATTENTION TO LIVELIHOOD RESTORATION

Resettlement includes two opposing sides that form a whole: displacement and livelihood restoration. The first side, displacement, is the process of people losing land, assets, or access to resources. The second side, livelihood restoration, is the process of assisting adversely affected people in their effort to restore and very preferably improve their previous living standards. All activities related to resettlement, from land recognition, land acquisition, compensation packages, to long-term community and regional development programmes, are aligned in time with these two processes, as illustrated below.

Cash compensation brings only temporary relief, if any, to affected people. It is a challenge in any society for most people to convert cash into sustainable measures for long-term social and economic benefits. For the financially inexperienced and vulnerable groups in societies affected by resettlement, this is even more difficult, particularly when productive capital that was a communal and family resource is converted to currency under the control of the head of household. It follows that resettlement is not only about compensation for resettlement losses, but it is also about social and economic development to re-establish the livelihoods of affected people. Successful resettlement programmes take advantage of opportunities to enhance the economic and social conditions of affected people. Cash compensation on its own should be the exception rather than the rule.

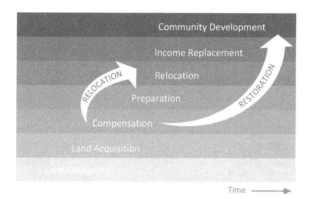

The issue of resettlement presents two opposing sides that form a whole: displacement and rehabilitation.

MANAGING RESETTLEMENT

We business leaders should see resettlement as a business risk and incorporate resettlement into our risk management framework. Visualising resettlement as simple, following a straight path from A to B, understates the reality involved. Unfortunately, the "line from A to B approach" to resettlement all too often characterises the mindset of technocratic personnel who are better equipped to embrace physical and financial project challenges, rather than the sustainable betterment of affected communities. To the engineer and accountant, the main interest, and often the only basis for financial rewards, are of a technical nature. Too often, resettlement is seen as an unwelcome burden to be dealt with by imposing engineering or accounting solutions.

We also need to put policies in place to guide and govern the planning and execution of resettlement, ensuring appropriate skills and commitment among the personnel responsible for implementation. In the absence of clear Company policy and real leadership, resettlement will never receive the priority it requires. Worse, resettlement may be viewed as a public relations exercise or as an impediment to the core business.

Resettlement is a management process with three phases: planning, execution, followed by monitoring and evaluation, as documented in a Land Acquisition and Resettlement Action Plan (LARAP), sometimes referred to as Resettlement Action Plan (RAP) or simply Resettlement Plan (RP). The main elements of an effective LARAP follow:

- Legislative Review – Understanding the legal framework for land acquisition is essential in developing plans for resettlement and livelihood restoration. It includes identification of all applicable national and regional laws, regulations, decrees, Government decisions, and similar provisions that apply to land acquisition, resettlement, and livelihood restoration.
- Thematic Maps – Developing thematic maps is helpful. These maps identify key site and regional features such as settlements, public infrastructure, soil types, vegetation cover, water resources, and land use patterns.
- Census – At its most basic, the census is a 100% sample, presented as an inventory of all households and businesses, formal and informal, affected by the loss of (or loss of access to) assets. To be useful, it must include additional information such as age, education, employment, and income, collected through the socioeconomic survey.
- Socioeconomic Survey – The socioeconomic survey is the basis for determining the socioeconomic conditions of individuals, households, and businesses that will be physically and economically displaced or otherwise affected by the project, essential for decisions on compensation and assistance.
- Asset Inventory and Valuation – The asset inventory is a process for registering all land and assets present in the affected area at the time and which need to be acquired for a project. This process is not unique to resettlement projects and is usually underpinned by a range of Government regulations.

- Defining Eligibility for Compensation and/or Assistance – Affected people eligible for compensation and assistance fall into three basic categories of (1) those with formal legal rights to land, including customary and traditional rights; (2) those with no legal right to land but present in the affected area at the time; and (3) those with no formal legal rights to land but having claims recognised under national laws or local customs.
- Stakeholder Engagement – Stakeholder engagement is a critical part of resettlement planning and implementation. The most important stakeholder in the process is the affected community, which needs to be involved in all phases of resettlement planning and implementation.
- Entitlement Measures – Eligible people, households, and businesses recorded by the census are typically entitled to compensation and assistance. Compensation can be monetary or in-kind, such as replacement land or provision of access to alternative land for animal grazing, or collecting forest produce.
- Livelihood restoration – Livelihood restoration comprises a set of measures implemented to address economic displacement – that is, to improve or, at a minimum, restore livelihoods and standards of living of affected people to predisplacement levels.
- Documentation – Two primary documents detail commitments to resettlement and livelihood restoration: a Resettlement Framework and a Resettlement Plan. The framework aims to establish resettlement principles and document commitments for adequately implementing land acquisition. The purpose of the plan is to present to all interested parties, including affected people, how the acquisition will be implemented, and by whom, as well as the time frame and resources.

ALLOCATING RESETTLEMENT BUDGET

Resettlement costs money, and the allocation of sufficient funds is essential. To be successful, resettlement requires a multi-year commitment. Land acquisition costs represent only a fraction of the total necessary resettlement budget. Cost items include expenditure for planning, community consultation and participation, capacity building, relocation, income restoration, and monitoring of resettlement success. Companies must, however, find the correct balance between the money spent on affected people and the money spent on personnel involved in planning and implementing resettlement. In more than one project, the irony is that money spent on consultant fees far outweighed the total expenditure on directly affected people.

Most companies are willing to spend more than the standard land value, recognising the inevitable inconveniences and stresses involved in resettlement. Unfortunately, this causes additional problems. First, it encourages land speculation, inflating land prices to the disadvantage of all except a few sellers. Second, it may lead to the aspiration of all local landholders to sell their properties. Those whose properties are not acquired may, in this situation, become active opponents of the project.

INVOLVEMENT OF COMMUNITIES IN LAND ACQUISITION AND RESETTLEMENT

Experience shows that mechanistic or paternalistic resettlement planning, while protecting people from immediate impoverishment, frequently fails to create a smooth transition to long-term livelihood restoration. In other words, resettlement cannot be planned in the same manner as physical infrastructure. The success of resettlement, like the success of most of the projects that cause it, will depend in part upon the involvement of those affected, as participation creates commitment and ownership. However, participatory processes can be time-consuming, expensive, and logistically cumbersome.

CUT-OFF DATE

The cut-off date is essential to ensuring ineligible persons do not take the opportunity to claim eligibility. Usually established at the time of the census, this sets the date after which people occupying the project area are not eligible for compensation and/or resettlement assistance. Similarly, fixed assets (such as built structures, crops, fruit trees, and woodlots) established after completion of the assets inventory, or an alternative mutually agreed on date, will not be compensated. However, establishing a cut-off date will help prevent but not eliminate an opportunistic rush of in-migration to the project area. The longer the time between the cut-off date and actual civil works, the greater the risk of opportunistic in-migration and associated claims.

WILLING BUYER/WILLING SELLER

People affected by a project may often be willing to sell their property and assets to the project owner voluntarily. Under such circumstances, it is essential to demonstrate that the transaction took place with the seller's informed consent; the seller was provided with fair compensation based on prevailing market values; and opportunities for the productive investment of the sales income exist.

We've decided to compromise. We keep the land, the mineral rights, natural resources, fishing and timber, and in return we'll acknowledge you as the traditional owners of it.

Source: Malcolm McGookin/CartoonStock

HOST COMMUNITIES

More often than not, the consequences of resettlement will impact two different communities; the settlement that is being relocated (displaced community) and the community receiving the newly relocated people (host community).

Consideration may also be given to more distant communities that have interacted socially or economically with the two directly affected communities. The host community may include the new settlements, or may be close to the new settlement so that there will be interaction between the host community and the resettled people. A business should engage in outreach activities to the host community to promote local integration and identify and address potential problems between host communities and resettled people. Provision of replacement land or access to alternative land for animal grazing or collecting forest products may reduce available resources for some members of the host community. Effective planning should minimise this impact. However, as resettlement is likely to require resource sharing, it is likely that the host community will blame the displaced communities and the project for any perceived economic losses and social conflicts that occur. Therefore, receiving the community's acceptance to host the new community is essential to ensure integration and sustain the resettlement. Developing an understanding in advance of the ethnic composition and social structure of both communities, as well as their previous history of interactions, is an obvious priority.

LAND MANAGEMENT SOLUTIONS

It is most important that companies involved in land acquisition and resettlement carry out a detailed and accurate assessment of landholders, maintain a consistent approach to negotiations, reach fair and just arrangements, and observe all their agreements and commitments. Land management solutions, mostly in the form of IT data management solutions, assist companies in managing their portfolios of land and mineral rights and responsibilities throughout the project life. These systems help companies to be vigilant about maintaining the obligations they have on land-related assets and their respective lease agreements and livelihood restoration commitments, while simultaneously ensuring regulatory compliance. The risk of failing to do so is clear; it can seriously damage a Company's reputation, its standing with host communities, and the bottom line.

QUESTIONS TO ASK

Business leaders should ask the right questions regarding resettlement. Many can and will arise, and unfortunately the following is probably not exhaustive, but it helps define the challenges we may face.

- Policy Framework – Do national policies and guidelines exist and are they adequate? Are institutional instruments in place to facilitate "land for land" compensation alternatives? Are there regional policies and regulations?
- Entitlements and Eligibility – Who will receive compensation and rehabilitation, and how much? How will these measures be structured to avoid or minimise cash compensation? Is recognition given to informal or customary land rights? How will we cope with land speculation?
- Social Preparation – Are the needs of women being taken into account? Will the needs of vulnerable people be met? Are neighbouring communities with a genuine interest in resettlement overlooked? Are potential sources of dissent and division in the community understood and offset?
- Community Participation – How best to consult with affected people? Does schedule allow for consultations that inch toward consensus? Will consultation lead to participation? Are host communities included in the consultation process?
- Organisational Capacity – Do the skills, staff, and organisational capacity exist to implement Resettlement Plans and assist in income restoration properly? Can affected people actively participate in resettlement?
- Role of Government – Do local Government institutions have adequate resources and capability to fulfil their roles? Will officials, perhaps in the guise of preventing land speculation or corruption, work to facilitate and profit from it?
- Conflicts of Interest – Do local joint-venture partners, employees, consultants, contractors, agents, intermediaries, brokers, appraisers, even translators and drivers, or their families, have possibly nonvisible interests in the local community and the acquisition-resettlement process?
- Criticism and Disinformation – Does the Company's stakeholder engagement programme include interaction with local, regional, national, and international mass media, social media, and Non-governmental Organisations? Is the Company prepared for surprises?
- Budget – How will land acquisition and resettlement be financed? Are allocated resources sufficient to fully cover resettlement, including reestablishment? Does the budget allow for a multi-year programme? Must host communities bear possibly

unexpected costs or does planning for community development benefits extend to host communities?

- Timeline – Is resettlement planning part of the early project planning, or is it more an afterthought? How does resettlement fit into the project implementation schedule? When will monitoring and evaluation of resettlement success start and end?
- Monitoring and Feedback – Is the grievance process adequate? Have internal and external audits been planned and scheduled? Who receives reporting, and how often?

Water resources in the future

Problems and solutions

Water may well be the world's most precious resource. Industrial operations can adversely affect water resources in two ways: by direct consumption, reducing the availability of water, and by contamination, rendering water unsuitable for other uses. In dozens of community consultations with which the authors have been involved, it is clear that the overwhelming issue of most concern among residents, faced with new resource development and where their land rights are not at risk, is the risk that local water resources will be damaged.

WATER SCARCITY

Water is also distributed unevenly around the globe: one-fifth of the world's population lives in areas of physical water scarcity, causing competition for limited resources. With less than half a per cent of the world's population, Canada has by some measures 20% of the freshwater. Some 69% of freshwater is now in ice caps and glaciers; as these melt, most of the water enters the oceans.

Furthermore, life in Earth's freshwater and marine ecosystems is not doing well. Society has been quite efficient at reaping the benefits of water, but this has come at a cost to the environment and the economy. For many centuries, water had the unfortunate task of transporting the pollution we emit. While significant strides have been achieved in the reduction of water pollution, more remains to be done.

Climate change and related shifts in weather patterns add to the pressure on water resources. Impacts of floods, droughts, and rising sea levels are predicted to intensify in the years ahead. Climate change is perhaps the most significant threat facing the global water system, simply because it involves many different risks.

DOI: 10.1201/9781003134008-9

Globally, water demand is predicted to exceed supply by 40% by 2030 (WRG 2019), with the inevitable result of increasing competition for water. Water scarcity and competing use will increasingly be a business risk that commands active management attention. For companies in water-intensive industries, implementing an effective water management strategy is essential to reduce the detrimental impact water scarcity may have on operations, Community Relations, and the bottom line.

Traditionally, we business leaders have approached water use from the production side by estimating how much water is used or polluted to provide goods and services produced. While the agricultural sector has by far the largest water footprint globally, industrial water use is often the dominant factor locally. Electricity generation is also a significant water consumer: it consumes more than five times as much water globally as domestic uses or industrial production.

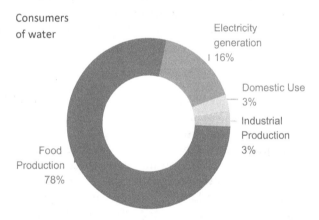

Based on data from Mekonnen et al. (2015) and Hoekstra and Mekonnen (2012).

However, an increase in water consumption is not necessarily a negative indicator of the state of the planet. Agricultural irrigation is crucial for us to be able to continue producing enough food. Water supply is also a prerequisite for most industrial operations. The challenge is that we do not have infinite amounts of water at our disposal. Water is a limited resource that we must protect, meaning that we cannot continue to allow ourselves to waste it, and must continually rethink how we obtain, use, and reuse it.

WATER RESOURCE PLANNING REQUIRES BASELINE DATA

In planning for an industrial development that could potentially affect water availability for other consumers, it is essential to evaluate the prevailing supply and demand situation, and the potential for additional supplies to be developed. This will enable the "safe yield" of surface water and groundwater sources to be assessed, as well as the risk that the safe yield could be exceeded during prolonged droughts.

A detailed understanding of existing water usage may also be valuable in the future to defend a Company against allegations that its operations have adversely affected other water users. If yields decline in a municipal water supply well or village wells run dry, the nearest industrial activity is likely to be blamed.

SAVING WATER

Saving water will never be a cliché. A direct link of water consumption to the environment, of course, is that water abstraction can stress the natural environment by altering natural ecosystems and draining aquifers. Water use has also a hidden environmental cost, as higher water consumption means more energy is used to treat and deliver water. Energy is also required to treat wastewater so that it can be safely returned to the environment or reused. Most energy production comes with an environmental cost. Thus, the more water we use, the more stress we are putting on the environment.

> The Earth, the air, the land, and the water are not an inheritance from our forefathers but on loan from our children. So we have to hand over to them at least as it was handed over to us.
>
> Gandhi

Rich countries can utilise sources of water that poorer countries cannot. For example, in wealthy countries, solutions such as the desalination of seawater can be an option. But this would be economically impossible in poorer countries, as it would be financially prohibitive to use desalinated seawater on a vast scale.

More efficient use of natural water resources is the first and most important step. This requires improved technologies with less water use and less wastewater generation. It is also necessary to clean and recycle water from production. A visionary water management strategy does not look at issues year-to-year. It considers the trajectory of business and products over the next decade and how operations can be managed to achieve established goals.

Industry understands that water is a valuable natural resource. That is why companies are continually improving their water management and conservation practices. Company efforts typically fall into three different categories.

- First, where possible, companies examine alternatives to using freshwater, or using it only once. Innovations in water treatment mean that companies can now effectively reuse water in their activities.
- Second, companies are diligent about capturing water produced during production processes, particularly in the resource sector. For example, in the oil and gas sector, companies recycle a large portion of water produced from exploration and production activities or reinject saline produced water back into geological formations.
- Finally, companies are building new water infrastructure systems to transport and deliver water more efficiently.

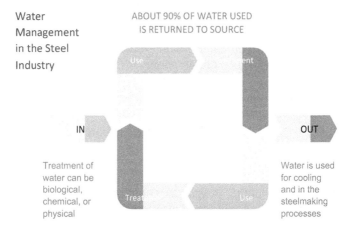

Water Management in the Steel Industry

ABOUT 90% OF WATER USED IS RETURNED TO SOURCE

IN

OUT

Treatment of water can be biological, chemical, or physical

Water is used for cooling and in the steelmaking processes

Source: worldsteel.org.

Saving water also involves the engagement of employees, local communities, and industry partners. As one example, Intel encourages employees to come forward with ideas on conserving resources and then matches the best ideas with a grant to develop these ideas. Experience demonstrates that most innovations that save water also save money, providing a payback in only a few years.

SMART WATER MANAGEMENT

Disclosing environmental data in annual Corporate Social Responsibility (CSR) reports is good, but offering real-time information is better. This means building a system that can provide real-time data about water usage and water discharges on a site-by-site basis, useable in day-to-day management.

A smart meter is an electronic device that records the consumption of production inputs (e.g., water or electric energy) in intervals of one hour or less. It communicates that information at least daily back to management for remote monitoring, reporting, and action. Smart meters also enable two-way communication between the meter and the central system, for example, to change the measurement interval.

With the development of network technology, especially the rise of the "Internet of Things," new technologies related to automatic meter reading and water management emerge in an endless stream: wireless network technology, network communication technology, data acquisition and management technology, including data security, expert systems, and so on.

A digital, or "smart," water network has sensors and Internet of Things devices that allow companies to manage water use more efficiently and effectively. Such a water network enables management to:

- Remotely monitor and identify problems (related to water quantity and quality), so that they can pre-emptively prioritise and identify maintenance issues.
- Provide information and tools to water users so that they can make informed choices about their water usage.
- Transparently and confidentially comply with regulations and policies on water quality and conservation.

Many mines and other industries have multiple water sources and may also have several wastewater discharges as well as recycling and reuse capabilities. Water production, storage, and consumption may change substantially over the life of an operation. For example, a mining project may be a significant consumer of water in its early years before becoming a net producer of water from mine dewatering. As oil and gas fields mature, the "water cut" percentage can increase dramatically, exceeding the capacity of reinjection or treatment facilities. Management of these inputs and outputs is facilitated by a water balance model that can be used to predict the consequences of different abstraction and discharge scenarios under a range of rainfall and evaporation conditions.

PROTECTING WATER

Industry leaders need to understand the many ways in which their operations can affect the water cycle:

- Agricultural Water Pollution in the form of eutrophication is when the environment becomes enriched with nutrients, potentially causing algal blooms disrupting normal ecosystem functioning, by using up all oxygen in water, blocking sunlight from photosynthetic marine plants, and producing toxins harmful to higher forms of life. As irrigation becomes increasingly important in a water-scarce world, its effects may be felt in aquifer depletion, soils waterlogging and salination, spread of waterborne diseases, and various types of pollution.
- Global Warming as an increase in water temperature due to climate change can result in the death of many aquatic organisms and disrupt many marine habitats, e.g., coral bleaching and poleward migration of key species.
- Atmospheric Water Pollution when water particles mix with sulphur dioxide, and nitrogen oxides, forming weak acid that precipitates as acid rain.
- Leakages from Underground Storage Tanks, a leading cause of soil and groundwater pollution.
- Oil Pollution in Water, from upstream oil and gas production, losses from storage facilities, spillage during delivery, or deliberate disposal of waste oil.
- Formation Waters produced by oil, gas, and geothermal operations may be saline or contain other pollutants of concern, even radioactivity.

- Nuclear Waste produced from industrial (nuclear power stations, nuclear-fuel reprocessing plants, and storage), or medical and scientific processes that use radioactive material. Mining and refining of uranium and thorium also generate nuclear waste. Radioactive substances can also be found in electronic waste containing medical equipment, instruments, cathode ray tubes, television sets, and x-ray machinery. Together with naturally occurring radioactive materials in geologic formations, ore bodies, and petroleum deposits, all these can be mobilised by human activity into freshwater resources.
- Sewage and Wastewater are sources of pollution in many lakes and rivers, especially in developing countries, as many people in these areas do not have access to sanitary conditions and clean water.
- Suspended Matter, as some pollutants do not dissolve in water. Runoff from disturbed land areas (e.g., clearing for agriculture or horticulture, construction, mining) can be contaminated by soil particles that eventually settle and degrade streambed habitats. Flow fluctuations may resuspend the sediments, transporting them further downstream (see changes to suspended sediment and downstream deposition below).
- Microbiological Water Pollution is a form of pollution caused by microorganisms such as bacteria, viruses, or protozoa. Serious diseases such as cholera come from microorganisms that live in water. Legionnaires' disease is a severe type of pneumonia caused by bacteria, called *Legionella*, that live in contaminated water from building water systems that are not adequately maintained.
- Oxygen Depleting Water Pollution by microorganisms feeding on discharges of biodegradable substances, using up the available oxygen and thus further degrading water quality.
- Biological Invasions when alien species (sometimes known as invasive species) from one region have been introduced into a different ecosystem where they were not previously present, often seriously disrupting ecological balance.
- Heat or Thermal Pollution is caused by water discharged from factories and power plants raising the receiving waters' temperature, reducing the dissolved oxygen content in water, with adverse effects on aquatic life.
- Changes to Levels of Suspended Sediment and Downstream Deposition – Suspended sediment levels can be generated

by stream bank or stream bed erosion or through sediment introduced by construction or land clearing activities. The opposite effect can occur when a dam is constructed for water supply or hydropower generation; sediment deposits on the reservoir bottom so that downstream water flows contain lower sediment concentrations. Both processes can have negative consequences.

- Waste Storage and Disposal – Waste stored in landfills or other storages may be eroded with contamination of surface waters. Rainfall percolating through a waste storage can leach soluble substances and so contaminate nearby groundwater.

- Mine Waste, Coal Stockpiles, and Acid Rock Drainage (ARD) – Mine wastes may be particularly problematic if the wastes contain sulphide minerals, a common situation in gold and base metal mines, and also in coal stockpiles and coal wastes. In a process known as ARD, sulphide minerals oxidise, generating acid solutions that can liberate heavy metals, leading to contamination of surface waters and groundwater.

- Acid Sulphate Soils, if drained, excavated, or exposed to air (e.g., by construction or lowering of the water table), can acidify water when sulphides react with oxygen, similar to ARD.

- Saltwater Intrusion caused by the abstraction of groundwater in coastal areas may cause the freshwater-saline water interface to migrate inland, contaminating groundwater supplies. A rising sea level can have the same effect.

- Accidental Water Pollution, ranging from minor oil spills to catastrophic water pollution events, e.g., the Sandoz Schweizerhalle chemical spill of 1986 after a fire caused a massive fish-kill in the Rhine near Basel; Exxon Valdez oil spill in Prince William Sound, Alaska, in 1989; the BP oil spill in the Gulf of Mexico in 2010; or Vale and BHP's Samarco iron ore tailings disaster in Mariana, Brazil, in 2015.

STANDARDS AND REGULATIONS

Point-source discharges are usually regulated by setting environmental standards in the form of effluent standards (ES) and ambient standards (AS). There are different existing philosophies in applying either just one of these standards or combinations of them for pollution management. ES encourage source control principles, such as effluent treatment and recycling technologies. AS requires considering the ambient response often associated with the concept of the "mixing zone," an allocated impact zone within which the numerical water quality standards can be exceeded.

ES are preferred from an administrative perspective because they are easy to prescribe and monitor (end-of-pipe sampling). However, from an ecological perspective, water quality control based on ES alone is limited, since it does not directly consider the quality response of the water body itself, or the cumulative impact of discharge over a prolonged period. AS are usually set as concentration values for pollutants that may not be exceeded in the water body. These have the advantage that they consider the physical, chemical, and biological response characteristics due to the discharge directly. They, therefore, put a direct responsibility on the discharger.

A combined approach includes the advantages of both water quality control mechanisms while largely avoiding their disadvantages. It follows that most recent regulatory attempts focus on defining plant- or site-specific mixing zones at the point of discharge. These zones consider the capacity of the receiving water to dilute the effluent and limit aquatic degradation spatially and temporally. While there is substantial variation in the specifics of the regulations worldwide, the combined approach shares two key elements: an ambient standard limit and a point of compliance expressed as a distance from the discharge.

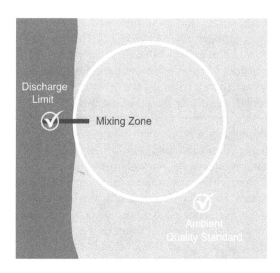

OFFSETS

Adverse impacts on water resources can be offset in several ways. One relatively common arrangement where the industry is being established in an underdeveloped area is for the industrial Company to develop a secure and safe water supply, not only for its own needs but for the local community. Similarly, the Company can contribute to the improvement of community sanitation, with benefits to previously contaminated water supplies. Potential offsets tend to be highly site-specific.

EMERGING CONTAMINANTS

In a broad sense, Contaminants of Emerging Concern (CECs) are any chemicals or microorganisms with potentially adverse ecological and human health effects, but which are not commonly monitored or regulated in the environment. CECs are not necessarily new chemicals; they may be substances that have been present in the environment for a long time, but whose presence and significance are only now being recognised (BP 12).

Emerging contaminants originate from a variety of product types, including human pharmaceuticals, veterinary medicines, illicit drugs, nanomaterials, personal care products, paints, and coatings. Many are used and released continuously into the environment, if only in small quantities. Some may cause chronic toxicity, endocrine disruption in humans and aquatic wildlife, and lead to bacterial pathogen resistance. Research done in the largest sewage treatment plant in New York City showed it is feasible to recover significant (possibly marketable) quantities of pure chemicals and even pharmaceuticals from the wastewater streams of megacities.

Microplastics, which are small pieces of plastic, less than 5 mm in length, are recognised as significant ocean water pollutants. Primary microplastics are less than 5 mm in size when they enter the environment. Secondary microplastics form with the breakdown of larger plastic products. Primary microplastics are present in various products, from cosmetics and personal care products to synthetic clothing fibres. Secondary particles form from plastic bags, bottles, and myriad other items; all these products readily enter the environment as waste. Secondary microplastics may be leached from plastic wastes disposed on land or break down in the aquatic environment.

Microplastics are not readily biodegradable. Thus, once in the environment, microplastics accumulate and persist. Microplastics have been found in more than 100 aquatic species, and they work their way up the food chains, from zooplankton and small fish to large marine predators. In a pilot study involving eight individuals from eight different countries, microplastics were recovered from stool samples of every participant (Parker 2018). In a recent study, microplastics were found in every one of the 47 human tissue samples examined from various parts of the body (Rolsky 2019). Even if, in the near future, all plastic waste could be intercepted before entering the sea, microplastics in water will persist as a chronic problem for generations. (In early 2021, the existence of microplastics as globally circulating air pollutants that move in and out of water bodies was recognised.)

SOURCE, SAVE, AND PROTECT

In most industrial projects, understanding flows of water into and from operations is relatively easy, but this is not so in mining. The water cycle of a mine site is interconnected with the general hydrologic cycle of a watershed. For groundwater, it involves the specific geologic and hydrogeologic structures intersected by the operation. Using and managing water at this scale carefully and responsibly requires a thorough understanding of the initial environmental conditions and forward-thinking leadership actions, best documented in a comprehensive strategic water management plan. The plan will provide detailed information about current water uses. It will also chart a course for water-efficiency improvements, conservation activities, and water-reduction goals, allocating resources to water-efficiency measures that offer the most significant beneficial impact. Besides planning, forward-thinking leadership actions include the following:

> Stewardship – Laws regulate water use and discharge in all countries. However, as industry leaders, we are expected not only to comply with the law but to demonstrate leadership in water use and management, particularly in the extractive industry where water plays an essential role in operations. Stewardship is our ethic to manage water resources responsibly. Taking a long view, what is legal today eventually may not prove to be ethical or moral.

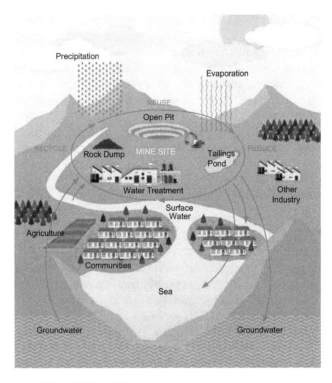

Based on ICMM 2012 – Water management in mining: A selection of case studies.

Appreciate the Value of Water – Water is a shared and often scarce resource. Understanding and addressing competing demands is a critical part of our water stewardship. Apart from managing impacts, we can often make a significant positive contribution to the provision of safe, clean, and adequate supplies of water to neighbouring communities. Most climate change projections indicate global water scarcity will continue to increase.

Consider Alternative Water Sources – Access to a secure and stable water supply is critical to most operations. The mining industry has often found innovative pathways to avoid competing with other water users and to use unconventional water supplies, such as seawater for mineral processing. Other operations harvest stormwater and focus on internal water reuse. As water scarcity increases on a warming planet, cost factors shift, and alternatives

such as condensing atmospheric moisture become more feasible.
Water accounting – A consistent approach to water accounting
across operations – including both quality and quantity – is the
first step to understanding a Company's needs and its water footprint. This can be challenging in the resources sector, given the
many activity streams within the typical site. Establishing a baseline against which future reductions can be measured is essential.
Is there a seasonal pattern to water use? Is water use increasing,
decreasing, or steady? What is causing major trends? Comparing
the total water supply baseline to water used by equipment and
applications will identify high water use activities and will allow
water-saving opportunities to be prioritised.

As is the case for energy consumption, the water balance
for most mining operations differs substantially during the
project life cycle. Consider a typical open pit mining operation.
Water supplies will be obtained initially from one or more surface water or groundwater sources. While the quantity of process water may be relatively constant throughout the project,
the water required for dust suppression increases as the mine
and its haulage system expand. In many open pits and most
underground mines, mining will eventually take place beneath
the water table, producing water that may partially or fully replace the primary sources, and excess mine water may, after
appropriate treatment, require discharge to the environment.
Meter, Measure, and Manage – Metering and measuring water
use helps assess opportunities for water use reductions. It also
identifies if the equipment is operated correctly and maintained
properly to prevent water wastage from leaks or malfunction-
ing mechanical equipment.
Assess Water-Efficiency Opportunities and Economics – Create a sustainability team with representatives from all departments that share responsibility for water use planning, usage,
and management. Ask the team to look into minimising water
losses during processing while maximising water recycling.
Think outside the box. In one example, a mining operation in a
water-scarce area took steps to reduce evaporative losses in the
heap leaching process. Another example is planting native and
drought-tolerant rehabilitation plant species to minimise the
need for supplemental irrigation.
Monitor Technological Innovation – Technology will continue
to be developed to find innovative solutions to the challenges

of sourcing water, reducing water demand, and designing more efficient and effective means of water management and treatment. The use of nanomembrane filtration technologies in effluent treatment is one example. Combined with a crystallisation unit, this can even achieve zero wastewater discharge.

Plan for Contingencies – Develop water emergency and drought contingency plans that describe how operations will meet minimum water needs during emergency, drought, or other water shortages and manage excess water during storm events. Assess the site for climate change risks, and reassess these periodically (BP 13).

Protect Water – Invest in ensuring that operations do not contaminate water. In addition to planned discharges, there are commonly various types of undesired or accidental water discharges: e.g., seepage or leakage from storage areas, tailings dams, and waste dumps; spillage of chemicals and fuels; and loss of containment due to natural events such as earthquakes or high rainfall events. Failure to manage these events could result in unacceptable consequences.

Management of Excess Water – Disposal of excess water involves different issues, even where the water quality is good. In arid areas, water produced by dewatering is usually discharged into ephemeral streams, creating "wetlands." While these may develop into productive ecosystems, they will wither and die following cessation of dewatering operations. Storage and evaporation involve expensive infrastructure and reinjection is technically complicated and expensive.

Engage with Stakeholders – Through stakeholder engagement we industry leaders can engage in constructive dialogue with communities and other water users about responsible water management, benefiting from different perspectives, and contributing to the debate. We are well-advised to establish a structured stakeholder engagement programme, which may include the community's participation in environmental monitoring (BP 07). Transparency – disclosure of water usage practices, beyond legal compliance, will work well to demonstrate the Company's commitment to responsible water stewardship. As is valid for energy use and GHG emissions, the Company should institute and meet targets for water use, working toward finding less water-intensive solutions.

Industry and biodiversity

The United Nations Convention on Biodiversity (1992) defined biodiversity as: "The variability among living organisms from all sources including, inter alia, terrestrial, marine and other aquatic ecosystems and the ecological complexes of which they are part; this includes diversity within species, and of ecosystems." It follows that biodiversity applies at different levels, namely: *ecosystems*, which may consist of a variety of *habitats*, supporting an assemblage of plant and animal *species*, each of which includes individual organisms with their own unique set of *genes*.

Any significant change in conditions, whether natural or human-made, can reduce biodiversity at one or more of these levels:

- Habitat loss, particularly the removal of natural vegetation for agriculture or urban development;
- Habitat degradation, such as from livestock grazing of natural pastures or wildfires;
- Invasive plants and plant pathogens displacing native species;
- Hunting and collecting of selected species for food, medicine, specimen, ornament, or household pet;
- Competition and predation from introduced animals; and
- Climate change, which particularly affects the distributions of species.

The public perception of biodiversity tends to be focused on so-called iconic species that are perceived to be at risk of extinction. Examples include rhinoceros, orangutans, polar bears, and koalas. Environmental activists and the media are similarly focused. However, the issue is much more complex and far-reaching than merely

DOI: 10.1201/9781003134008-10

protecting well-known iconic (sometimes called "charismatic") species.

Biodiversity has recently become a major environmental issue. However, it has a rather long history. A century and a half ago, Charles Darwin identified the first principle of modern biodiversity: all species are linked to a single great tree of life. All can be traced back to a presumed single original species at some distant time in the geological past. Darwin and his contemporary Alfred Russel Wallace developed evolution theories based on natural selection, where species developed in response to their natural environments. The most adaptable proved to be the most successful – survival of the fittest – a process driven by diversification, and at times, extinction of species.

Since Darwin presented his theory, biodiversity on Earth has decreased significantly, in a process that reduces the adaptability and sustainability of life on the planet. Of course, this is not the first time there have been massive reductions in the number of species. Five major extinction events have extended from the Ordovician-Silurian extinction some 444 million years ago to the Cretaceous-Paleogene about 66 million years in the past. The greatest was the Permian-Triassic "Great Dying," 250 million years ago, the single worst event ever experienced by life on Earth. Over a geologically brief 60 millennia, 96% of marine species and 75% of land species became extinct, while the world's forests were wiped out for 10 million years (nationalgeographic.com).

At present, many biologists believe that the rate of species loss is greater than at any time in human history. They believe plant and animal species are being lost one hundred times faster than background extinction rates. While previous extinctions were caused by combinations of ice ages, volcanism, massive increases in greenhouse gases, and asteroid impact, the current event is blamed on a single species, *Homo sapiens*, exerting its dominance over the planet. No surprise then that the geological epoch dating from the commencement of significant human impact on Earth's geology and ecosystems sometimes is now referred to as the *Anthropocene* – named after us.

Extinction of species may not always be a bad thing. Data suggest that humans, by 1977, managed to eliminate *Orthopoxvirus variola*, the smallpox virus, outside a small number of laboratory stocks. This removal of an age-old scourge on humanity was accomplished by using its cousin, *O. vaccinia*, to develop population immunity, as no treatment or cure was ever found.

This discovery is credited for creating the medical science of vaccination. *Enterovirus C*, the poliovirus, is now restricted to a few small geographic regions; it might already have been eradicated had not terrorist groups repeatedly attacked public health workers in Nigeria and Afghanistan. In 2011, *Rinderpest morbillivirus*, which caused a severe cattle disease, was officially declared eradicated. After smallpox, Rinderpest is only the second disease in history to be eliminated and is the first non-human virus extinction.

The ongoing global eradication campaign for mosquito species that spread diseases also seems to be morally and economically justified. Should humankind be able to destroy the dozen or so worst disease vector mosquitos, some 3,500 other species in the *Culicidae* Family would remain to fill their places in the food chain. But what about the many other species that contribute to Earth's biodiversity and thus its ability to adapt to change? We know that biodiversity loss reduces the natural resources available to us and makes ecosystems more fragile and less productive. Clearly, some species are more important than others with apex predators playing a particularly important role.

The International Union for Conservation of Nature (IUCN) produces the IUCN Red List, which has assessed the status of around 160,000 plant and animal species in terms of the following categories:

- Data Deficient (DD) – species for which abundance and distribution data are lacking;
- Least Concern (LC) – species that are not categorised as NT, WU, EN, or CR;
- Near Threatened (NT) – species that are close to qualifying or likely to qualify for a threatened category in the near future;
- Vulnerable (VU) – species facing a high risk of extinction in the wild;
- Endangered (EN) – species facing a very high risk of extinction in the wild;
- Critically Endangered (CR) – species facing an extremely high risk of extinction in the wild;
- Extinct in the Wild (EW) – species no longer present in the wild, based on extensive field surveys;
- Extinct (EX) – where there is no reasonable doubt that the last individual has died.

In addition to this categorisation, the Red List provides information for each species about range, population size, habitat and ecology, use and/or trade, and threats.

Industry in the broad sense poses numerous threats to biodiversity conservation, either directly from habitat loss due to agriculture and extractive industries or indirectly through species lost from pollution and climate change, most notably due to combustion processes in the energy and transportation sectors.

While biodiversity issues today attract considerable academic, activist, and media attention, the concepts are often too abstract for business leaders to incorporate into business planning, let alone develop strategies for investing in biodiversity protection and the opportunities deriving from it. Interest in this is certainly lagging behind the awareness of opportunities now being perceived in responding to climate change. Business leaders have long understood nature, whether in terms of biodiversity or other natural resources, as common public goods to be diverted through rent-seeking to be exploited for immediate returns on investment. There also seems to be a lack of useful knowledge on the subject in a language comprehensible to most industry leaders. It can reasonably be predicted, as with climate mitigation and adaptation activities, that biodiversity preservation and enhancement investments will in the future create value, and thus profits, for innovative entrepreneurs.

Business and biodiversity guidelines do exist. The Equator Principles (EPs) are internationally considered the "gold standard" for sustainable project finance (BP 04). They are entirely based on the International Finance Corporation (IFC) Performance Standards on Social and Environmental Sustainability (notably PS 6 as described below) and the World Bank Group's Environmental, Health and Safety General Guidelines and are the premier example. With the growing ESG role in business, the EPs make a strong case for business leaders to understand their responsibilities in conserving biodiversity and ecosystem services and even in exploring opportunities for private investment in this domain.

CRITICAL HABITATS AND BIODIVERSITY IMPACTS

Whether or not biodiversity conservation applies to a particular project development is established during the environmental risks and impacts identification process. Biodiversity conservation requirements are primarily applied to projects located in Critical Habitats, which requires an understanding of exactly what defines a Critical Habitat. Critical Habitat, as defined by the IFC, was introduced in 2012 in its Performance Standard 6 (PS 6) on Biodiversity Conservation and Sustainable Management of Living Resources. The Critical Habitat delineation is designed to identify high biodiversity value areas across the world where development would be particularly sensitive and require special attention. The concept considers global and national biodiversity conservation priorities and was developed in consultation with many international conservation organisations. It also considers many pre-existing conservation value classifications, such as Key Biodiversity Areas, Important Bird Areas, and Alliance for Zero Extinction Sites, as well as the more established Protected Area Management Categories, World Heritage Sites, and Ramsar Wetlands. The text of PS 6 and its Guidance Note, and all of IFC's follow-up publications, are readily available and (along with publications of various non-governmental organisations (NGOs) and industry groups) go a long way toward developing the "...language comprehensible to most industry leaders." Hopefully, this chapter can also contribute to this effort.

Critical Habitat is established where a habitat site or area holds biodiversity values meeting any of the following five criteria:

- Habitat of significant importance to IUCN Red List Critically Endangered (CR) or Endangered species (EN).
- Habitat of significant importance to endemic and/or restricted-range species.
- Habitat supporting globally significant concentrations of migratory species and/or congregatory species.
- Highly threatened and/or unique ecosystems.
- Areas associated with critical evolutionary processes.

The presence of Critical Habitat does not explicitly exclude development. Instead, it sets higher conservation goals when assessing developments through the mitigation hierarchy. Where Critical Habitats are identified, developments impacting on Critical Habitats must demonstrate a "net gain" in biodiversity values through mitigation and offsetting programmes.

Biodiversity is not distributed evenly on Earth and Critical Habitats are more common in the tropics. Tropical forest ecosystems cover less than 10% of Earth's surface but contain about 90%

of the Earth's species. It was not by chance Darwin and Wallace carried out their original field studies in the tropics. However, Critical Habitat can exist anywhere, even in unlikely places, as the following case study shows.

Source: Henrik_L/ stockphoto

Just a few centimetres tall, it smells a bit strange (German entomologists call it "Aprikosenkäfer" because it smells like apricots), and lives in mouldy, wet tree hollows, which it rarely leaves: The Hermit (*Osmoderma eremita*) is probably the rarest large beetle of Germany. As a so-called jungle relic, the species lives in Stuttgart Castle Park in ancient trees, many of which had to be felled for the ambitious Stuttgart 21 Railway Project.

In Germany, the once widely distributed insect now only survives in tiny populations in small islands, such as the few hollow trees in Stuttgart Castle Park, a Critical Habitat through a single criterion: the presence of a highly threatened and unique ecosystem in the form of ancient hollow but still living, standing trees.

Local extinction of a population of the rare beetles could be caused by felling this small number of trees.

The occurrence of this Hermit in Stuttgart Castle Park not only delayed the Stuttgart 21 construction project, but additional environmental regulations designed to protect biodiversity contributed to the costs spiralling out of control. To add to the controversy, the developer accused project opponents of placing "fake beetle marks."

Major changes in land use, of agricultural land, in forests, and in urban areas took place throughout Europe over recent decades, causing remaining stands of large old-growth free-standing trees to become Critical Habitat.

Sub-surface organisms such as stygofauna and troglofauna may also present challenges, due mainly to the lack of knowledge concerning their distributions and habitat requirements. Mining project approvals in Western Australia have been affected by discovery of such organisms.

Climate change also creates unique or unusual habitats. Permanently moist areas in otherwise arid environments are one example. These can support remnant populations of plants and animals that were widespread during different, usually wetter periods, but they have contracted to one or more refuge areas as the climate has changed. Examples are ferns confined to permanently shaded areas at the bases of cliffs in central Australia.

These relict species may expand their distributions again when cooler or wetter conditions return. Similarly, the rare beetles in the Stuttgart Castle Park may also expand their distributions when conservation efforts create new suitable habitats.

MITIGATION HIERARCHY TO AVOID AND MINIMISE BIODIVERSITY LOSS

Biodiversity conservation efforts refer extensively to the mitigation hierarchy, a series of sequential steps taken throughout the project life cycle to avoid or limit negative impacts on biodiversity and to achieve no overall negative impact on biodiversity, or on balance a net gain (also referred to as No Net Loss and the Net Positive Approach).

Avoidance is the first step of the mitigation hierarchy. It comprises measures taken to avoid biodiversity impacts from the outset, such as selecting alternative locations or careful spatial or temporal placement of infrastructure or disturbances. A resource footprint is fixed, but buildings and infrastructure positioning may have flexibility. Avoidance is often the easiest, cheapest, and most effective way to reduce potential adverse impacts, but it requires biodiversity to be considered in the early stages of project development and engineering design.

Minimisation involves measures to reduce the duration, intensity, and extent of direct and indirect impacts that cannot be wholly avoided. Examples include reducing noise and pollution, designing infrastructure lighting not to disturb wildlife, or building wildlife crossings over and under roads. Indirect impacts must be carefully and strictly managed, as these can sometimes have more significant effects than anticipated. Restricting public access to project roads that enter forest areas through portals and security patrols is very important and must be managed well. Otherwise, Company roads provide improved access for the hunting or capture of wildlife, removal of other forest products, and even agricultural clearings. This goes beyond simply setting guards and checkpoints, as it requires managing relations with local communities and regional economic interests.

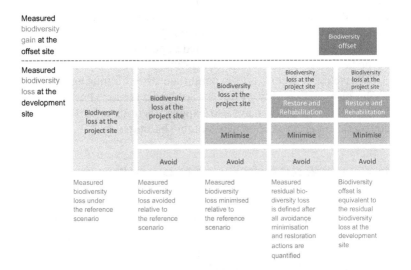

Source: Adapted from Rio Tinto (2012). Rio Tinto and Biodiversity: Working Towards Net Positive impact, Rio Tinto PLC London, UK, Rio Tinto Limited, Melbourne Australia. Available at www.riotinto.com/ourcommitment/features-2932_8529.aspx

Rehabilitation/Restoration involves measures to improve degraded ecosystems or restore removed ecosystems following exposure to impacts that cannot be completely avoided or minimised. Restoration aims to return an area to the original ecosystem that existed before impacts. Rehabilitation only seeks to restore essential ecological functions and ecosystem services (e.g., planting trees to stabilise bare soil). Rehabilitation and restoration are frequently needed toward the end of a project's lifecycle. In some cases, they may be possible during construction and operation (e.g., after temporary borrow pits have fulfilled their purpose, or when revegetating road batters after construction).

Collectively, avoidance, minimisation, rehabilitation, and restoration serve to reduce, as far as possible, the residual impacts a project has on biodiversity. Typically, however, additional steps in the form of offsets will be required to achieve no overall negative impact or a net gain for biodiversity.

BIODIVERSITY OFFSETS TO ACHIEVE NO NET LOSS

As with all environmental management efforts, business leaders should ensure that the above mitigation hierarchy framework is rigorously applied in their operations and projects, a widely accepted approach for biodiversity conservation that meets the standard for "best practice." In addition to prevention and mitigation measures, biodiversity offsets are conservation outcomes serving as last resort measures to compensate for adverse and unavoidable residual biodiversity impacts. Offsets aim to achieve *No Net Loss* (NNL) and preferably a *Net Gain* (NG) of biodiversity when projects take place. The achievement of NNL or NG is dependent on measurable, appropriately implemented, monitored, evaluated, and enforced offset schemes, such as protecting threatened forests or restoring wetlands.

In many jurisdictions, biodiversity offsets are not optional. First used in the USA in the 1970s to mitigate damage to wetlands, many countries have now adopted biodiversity offset programmes. In more than 100 countries, laws or policies are in place that require or enable the use of biodiversity offsets, or their use is currently being considered. Depending on the jurisdiction, various schemes for compensating biodiversity impacts may be available. One-off offsets are carried out directly by the developer (or an appointed contractor) once predicted residual adverse impacts had been evaluated. The developer assumes financial and legal liability for implementing the offsets. Verification is generally undertaken by a Government agency or an accredited third party. In some jurisdictions, a Government agency may set an in-lieu fee that the developer must pay to a third party to compensate for residual adverse biodiversity impacts. The third party (i.e., the offset provider) assumes the further financial and legal responsibility.

Biobanking is yet another option. Once predicted adverse impacts are evaluated, the developer can purchase offsets directly from a public or private biobank. Biobank refers to a repository of existing offset credits; each credit represents a quantified gain in biodiversity resulting from actions to restore, establish, enhance, or preserve biodiversity (e.g., wetlands, stream, habitat, species). As under the in-lieu fee arrangement, financial and legal liability are transferred from the developer to the provider.

Determining the size and type of biodiversity offset is often complicated and requires specialist expertise in evaluation, design, and implementation. In many developing countries, offsets are much easier to design than implement, as regulatory mechanisms are not in place. Accordingly, attention to earlier steps in the mitigation hierarchy is essential to ensure residual impacts are minimised as much as practically feasible. The specific environmental and social context and regulatory system within which a project is located will have a significant bearing on the types of offsets that are suitable and feasible. The best biodiversity offsets should benefit local communities as well as biodiversity values. The success of biodiversity offsets is thus very context- and location-specific.

ADDITIONAL CONSERVATION ACTIONS

Additional Conservation Actions are extra measures with positive – but often difficult to quantify – effects on biodiversity. These qualitative outcomes do not fit easily into the mitigation hierarchy but may provide crucial support for mitigation actions. For example, partnerships with conservation NGOs for species recovery programmes may assist in minimising impacts. Several mining companies have entered into long-term partnerships with international conservation NGOs, such as The Nature Conservancy, Conservation International, Fauna and Flora International, and their country programmes.

Partnerships to implement community forestry programmes with local communities are another example and may reduce adverse effects on the broader landscape and society. Awareness activities may encourage changes in Government policy that are necessary for the implementation of novel mitigation strategies. Research on threatened and economically important species may be essential for designing effective minimisation measures. Capacity building for local stakeholders may help to engage villagers with biodiversity offset implementation.

Demonstrating a corporate and asset/project-level commitment to understanding and managing biodiversity risks is essential and is achieved through a hierarchical management system. It commences with corporate commitments to sustainability and moves down through corporate environmental, biodiversity, and sustainable development standards to an asset or project-specific biodiversity action plan or strategy. This serves as the basis for operational biodiversity management plans and procedures specific to project lifecycle phases, such as exploration, construction, operations, and closure.

Corporate Commitment to Sustainability

Corporate Standard or Policy — Corporate Level Environment or biodiversity standards which must be complied with all assets.

Asset /Project-level Biodiversity Action Plan or Strategy — Biodiversity Action Plan or Strategy which is overarching demonstration of an asset's or project's, understanding of the biodiversity values and threats and the approach to application of mitigation hierarchy.

Asset /Project-level Biodiversity Management Plan & Procedures

Site-specific Construction, Exploration, Project Management Plans, and Procedures — Action/Project Level Biodiversity Management document which are fit for purpose, implementable on the ground plans and procedures

Site/Project Phase specific management plans and procedures may be incorporated info environmental plans and procedures

Source: Adapted from Temple, H.J., Anstee, S., Ekstrom, J., Pilgrim, J.D., Rabenantoandro, J., Ramanamanjato, J.-B., Randriatafika, F. & Vincelette, M. (2012). Forecasting the path towards a Net Positive Impact on biodiversity for Rio Tinto QMM. Gland, Switzerland: IUCN. x + 78 pp.

ALIEN INVASIVE SPECIES

Humans frequently introduce species to new habitats, in complex combinations of intentional, inadvertent, and sometimes inexplicable events. Introduced plants and animals often pose much greater biodiversity threats than the direct effects of commerce or industry. Awareness of species migrations only develops after a previously balanced ecosystem is suddenly disrupted, and local species start being eaten or displaced. Business leaders need to know how their operations could trigger such events or may need to deal with their existence in a site region.

Ballast water disposal, an obscure issue to most people, has disrupted marine ecology worldwide. Of the massive modern increase in global trade volume, at least 80% is carried on ships. And seaborne trade is inherently unbalanced – bulk carrier ships move commodities like grain, ore, and fuels to markets and often return empty. Stabilising global trade flows may be a complicated matter for Government economists but stabilising the ships is simply a matter of flooding their ballast tanks with seawater. And in that process, thousands of types of hitchhikers – microbial pathogens, algae, fish, arthropods, molluscs, eggs, larvae, and a taxonomic jungle of worms, sponges, jellyfish, and "parasites" amounting to around 7,700 known problem species – may be taken aboard and transported to a distant destination. Like any stowaways, they are not a problem until their discovery after arrival. Often this requires expensive measures to protect native aquatic life, fisheries, infrastructure, and sometimes public health.

A weed or a pest is any organism, however harmless or useful in its natural habitat, that becomes a significant and usually expensive problem when it shows up where it is not wanted. "Invasive" means just that; the organism invades a landscape or waterbody and often changes it permanently. But the accidental migrations have not been the only problem, or even much of it; such imports and exports can be at least somewhat regulated. Many of the worst weeds and pests throughout the world were deliberately carried between countries and continents as ornamental plants, food, pets, or for sport or recreation. Some of the most severe invasives were introduced to control other weeds or pests, usually unsuccessfully.

Generations of mine rehabilitation were carried out with the focused goal of making disturbed land "green" again as rapidly as possible, using the fastest-growing available cover crops and trees. Unfortunately, fast-growing is often nearly synonymous with invasive. The use of non-native species is far less common now than in the past, but only as rehabilitation has ceased to be a support activity to become a core mining operation, due to increasing environmental consciousness. Inadvertent weed introduction has occurred through seed contaminated with weed seeds, even in commercial seed supplies. Rehabilitation plantings, both cover crops and tree seedlings, need to be carefully trialled and closely monitored for signs of unwanted species. Weeds and plant pathogens can also be introduced by being on or in items relocated from other areas. Many mining projects employ plant and equipment items previously used on other sites. Soil pathogens, insects, and fungi can be introduced from these items unless stringent preventive measures are applied. Similarly, it is common for construction contractors to relocate workers' accommodations from one construction site to another. In one such case, numerous Redback spiders were found when portable dongas (accommodation buildings) from Australia were being unloaded at a mine site in Indonesia. This venomous spider (*Latrodectus hasselti*), common in Australia, does not occur in Indonesia. Fortunately, the infestation was discovered in time to eradicate the spiders before they could spread. Giant African land snails (*Achatina fulica*) invaded Lihir Island after the gold mine was established there, probably arriving with timber imported by Island residents, and seriously threatened the island agriculture. This may not have occurred without the wealth created by the project; the mining Company is leading efforts to eradicate the molluscs, a "hand-picking" process that moves at a snail's pace, so to speak.

While not all introduced species become invasive, numerous domesticated animals and household pets have gone feral (wild) creating a wide range of problems. Dogs, cats, and other pets kill native birds and small animals. Feral goats have changed the vegetation cover on whole islands in the Pacific and Caribbean. Feral pigs (*Sus scrofa*, with 19 subspecies) damage crops and threaten humans in some 130 countries; with a warming climate, they are rapidly expanding their range in North America and worldwide, only having been (probably) eradicated in a few countries.

Such examples abound; another hundred or so could easily be added. Once again, the key conclusion is that one doesn't know what one doesn't know, and such "Unknown Unknowns" often prove to be important. Business leaders must thoroughly discard any attitude that considers a rigorous environmental assessment to be merely a regulatory hurdle or a public relations exercise rather than a critical planning and risk management tool and focus on identifying the best environmental consultants and specifying a thorough scope of study.

BIOTECHNOLOGY AND GENETIC RESOURCES

In closing this discussion of biodiversity issues, it is instructive to consider the revolution in biotechnology that is changing the world. If the importance of the subject is not immediately obvious, it is illustrated by recent events affecting the entire planet. In the second week of the 2020 pandemic year, before the Government of China was ready to admit SARS-2 coronavirus could be transmitted "human-to-human," it posted the viral genetic sequence on the Internet. While Chinese doctors were still learning that the virus was not simply causing pneumonia but mounting an all-body attack as they watched on scanners, rapidly creating blood clots that destroyed lungs and other organs, laboratories all over the world were simultaneously creating diagnostic methods to detect the disease and vaccines to prevent it, based entirely on digital data. Actual specimens of the virus were never shared – but soon enough every country could obtain them locally.

If informatics has learned that data itself is a resource to be exploited by the tech industry, then through "bioinformatics" the biotech industry knows data contained in cellular, noncellular, and acellular organisms' DNA and RNA are also resources of great value for entrepreneurs who develop opportunities to exploit them. This will increasingly hold true for everything from viruses to whales and is why the current rate of species loss is of much more than academic concern. No one questioned why funds appeared overnight for the effort to drive extinct the *Sarbecovirus SARS-related coronavirus 2*, while it was holding the entire world economy hostage and beginning to evolve rapidly even more dangerous strains. But no one knows how many of the 20 or 50 or 150 plant or animal species going extinct every day from human activity or climate change might contain genetic information useful to humans, and there is minimal funding to learn – or even to obtain samples of the DNA being lost.

Minimal does not equate to nothing, but to a chronic shortage of investment. Efforts to preserve genetic resources do exist (e.g., in the form of IFC PS 6); the issue was clearly recognised in the UN Convention on Biological Diversity (CBD) of 1992. The CBD Preamble recognises the "intrinsic value of biological diversity and its components," as well as its economic value. It affirmed States' sovereign rights over their own biological resources and recognised Indigenous and local communities following traditional lifestyles are dependent on biological resources. The CBD expressed the desirability of equitable sharing of benefits arising from the use of traditional knowledge, innovations, and practices relevant to the conservation of biological diversity and sustainable use of its components. CBD acknowledged substantial investments are required to conserve biological diversity and achieve a broad range of environmental, economic, and social benefits from those investments. Thus, the Convention focused attention on Access and Benefit Sharing of genetic resources and the technologies to exploit them.

As has been discussed at length in a recent book (Lawson and Adhikari 2018), the CBD's initial focus on obligations to collect physical-biological material has evolved into complex disputes and challenges about how Access and Benefit Sharing should be implemented and enforced. These include: repatriation of resources, technology transfer, Genetic Resources related to Traditional Knowledge (GRTK) issues and cultural expressions, open access to information and knowledge, farmers' rights, naming conventions, and schemes for accessing pandemic virus specimens and sharing their DNA sequences. As Lawson and Adhikari note: "Unfortunately, most of this debate is now crystallized into apparently intractable discussions such as implementing the certificates of origin, recognizing traditional knowledge and traditional cultural expression as a form of intellectual property, and sovereignty for Indigenous peoples."

Indonesia is an example of a country that is well aware of this issue. It passed Law No. 5 of 1994 ratifying the CBD in the same year as Law No. 7 ratifying the agreement to establish the World Trade Organization, including the TRIPs Agreement (Trade-Related Aspects of Intellectual Property Rights) for regulating Intellectual Property. The 2020 Patent Law is considered to be relatively liberal, harmonising with CBD and TRIPs and allowing genes and even organisms to be patented (Barizah 2020). Considering terrestrial and marine life together, Indonesia is probably the most biodiverse nation on the planet. Trade in its biological resources brought in foreign interests over a millennium, including more than four centuries of colonialism. Since the 1970s, it has struggled to entice either foreign or domestic investment in agriculture; what success has been achieved was virtually all in oil palm, with some pulpwood plantations.

In recent years, the Agriculture Ministry's Research and Development Agency (BALITBANGTAN) established an active genetic resources collection and registration programme, recognising the nation is impoverished in this area through inability to adequately document Indonesia's rich diversity. From 2004 to 2014, the Agency listed 1,548 varieties and granted rights protection to 328 plant varieties including cereals, plantation/industrial commodities, ornamentals, vegetables, fruits, tubers, livestock feeds, nuts/pulses, and spices-medicinals as well as some livestock varieties. By the end of this period, BALITBANGTAN operated 12 "collection plantations," since increased to 45 (Website of Agriculture Ministry Secretary General, accessed December 2020).

Some of the most exciting business opportunities in the decades ahead will be in slowing and reversing climate change and biodiversity loss, for those who can find and develop such opportunities. For the rest of us, we must acknowledge extinction of plant and animal species will inevitably continue, while opening wildlands that are their last remaining habitats will continue to bring humanity into contact with new pathogens and parasites and invasives. Businesses operating in proximity to these processes must ensure they are not their causal agents.

Enterprise risk management

Putting ESG risks into a business context

Enterprise Risk Management (ERM) is a relatively new and still evolving discipline receiving increasing attention from the Board and senior management. There is a wide range of views on how ERM should be implemented and what constitute best practices, but most interested parties would agree ERM is a systematic process applied within an enterprise designed to manage significant risks to business activities, outcomes, and long-term success. BP 11 provides an overview of critical aspects of ERM, particularly determining unacceptable risk. Those seeking more detailed information could refer "A Risk Practitioners Guide to ISO 31000" (IRM 2018).

Risk assessment to inform decision-making has been standard practice in industry for many years. However, the need for risk-based decision-making has never been greater, given rapid social and technological change, geopolitical uncertainties, and more stringent compliance requirements.

> You can't build a reputation on what you're going to do.
>
> Henry Ford

A well-implemented ERM programme can deliver a range of benefits: reducing the likelihood and consequence of incidents, better informed strategic decisions, increased operational efficiency, competitive advantage, and reputational benefits, to name a few. It is important to recognise the various types or classes of risk that an organisation may face, as the kind of risk will largely determine the controls and resources required for its management.

DOI: 10.1201/9781003134008-11

Top 10 Enterprise Risks by Sector in 2021

	Mining and Metals Companies Published by Ernst & Young Global Limited		Oil and Gas Companies Published by consulting firm BDO USA
Strategic Risk	Poor formulation and execution of business strategy, including failed Joint Venture or partnership arrangements, can risk a range of impacts, ranging from under-performance to business closure.		
Financial Risk	Financial risk generally falls into three categories: liquidity risk, capital availability risk, and capital structure risk. Additional financial risk may arise from changes in foreign exchange and interest rates and hedging strategy.	Capital agenda, e.g., appetite for risk and approach to capital allocation must ensure new opportunities are not missed.	Volatile oil and gas prices and Regulatory and legislative changes (equal placing). Inaccurate reserve estimates. Inadequate liquidity or access to capital. Inadequate or unavailable insurance cover.
Organisational Risk	Factors such as the performance, retention and availability of employees, ineffective leadership, agency risks, organisational costs, and cultural alignment.	Workforce, e.g., change in the corporate culture of mining and metals companies due to the pandemic.	

(Continued)

Top 10 Enterprise Risks by Sector in 2021

Operational Risk	Factors that prevent or impair normal operations and the achievement of targets include safety incidents, unplanned shutdowns, low recoveries, inadequate orebody knowledge, geotechnical hazards, loss of data, disruption to logistics, and cost of raw materials.	Productivity and rising costs, i.e., due to disrupted supply and the impact of economic uncertainty on demand.	Inability to expand or replace reserves. Operational hazards.
External Risk	External risks range from reduced market demand to changing competitive behaviour to limitations on capital availability. Structural or cyclical change within the industry can create high-risk situations.	High-impact risks, e.g., pandemics. Decarbonisation and green agenda. Volatility. Digital and data innovation.	U.S. general economic concerns. General industry competition.
Compliance Risk	Compliance risk may involve breaches of laws, regulations, operating permits, and industry codes.	Licence to operate (LTO), considered the no. 1 issue for miners.	Environmental restrictions and regulations.
Country Risk	Risk associated with a particular country or region may involve corruption, lack of law and order,	Geopolitics, reflecting the shifting balance of power among the world's largest economies.	

(Continued)

Top 10 Enterprise Risks by Sector in 2021

	weak judicial system, inadequate basic services, and political instability.	
Hazardous Risk	Hazardous risks cause direct harm to people, the natural environment, and property. They include natural disasters, pandemics, pollution, and accidents.	Natural disasters and extreme weather.
Reputational Risk	Reputation is sometimes included as a class of risk. However, a reputational loss can be a consequence of the mismanagement of any enterprise risk.	

Caution must be applied in the application of such lists to enterprise risk assessment at a Company level. Site and country factors strongly influence many of the risks faced by individual companies. Also, with due respect to the accounting/management firms that produce such lists, they often appear to underestimate operational risks. Site closure liability represents a significant risk for many mining companies but is missing from the listing above. Curiously, the extensive discussion that accompanied the above listing also made no mention of tailings facility risk, undoubtedly one of the greatest risks associated with the modern mining industry.

International standards organisations have developed various ERM frameworks, primarily addressing risk strategy and management frameworks rather than the operational or "nuts and bolts" aspects of risk management. The two most widely referenced are International Standardization Organization, ISO 31000 (2018) and Committee of Sponsoring Organizations (of the Treadway Commission), COSO (2017), developed by organisations with very different professional backgrounds and differing in key aspects.

Key Differences between ISO 31000 and COSO Risk Management Standards

ASPECT	ISO 31000 (2018)	COSO (2017)
Structure	More management system oriented (16 pages in length).	More detailed and prescriptive (over 100 pages in length).
Approach	Greater emphasis on supporting organisations in achieving goals rather than simply avoiding risk.	Viewed by some as being overly focused on avoiding risk and failure.
Development	Development included submissions from over 70 countries.	Developed in partnership with accounting firm PwC.

Aspect	ISO 31000 (2018)	COSO (2017)
Target Audience	Risk management practitioners, educators, etc.	Accounting and auditing professionals.
Scope	Risk management and strategic planning.	Corporate governance aspects in addition to risk management.

Neither ISO 31000 nor COSO was designed to support auditing or certification of companies. They are best viewed as technical guidance for developing a risk management framework. ISO 31000 is better suited to the needs of a resources Company already familiar with management systems such as ISO 14001.

POLITICAL RISK

Substantial political risks are involved for a Company investing in an industrial project outside its home country. Mining enterprises are peculiarly challenged by political risks for several reasons:

Firstly, much of the world's most attractive mineral wealth resides in jurisdictions that are under-explored precisely because political conditions have historically discouraged foreign investment. Secondly, mining businesses are obviously captive to the location of the orebodies and can't relocate after an investment decision is taken. Thirdly, mining is invariably capital intensive, and most capital must be sunk at the start of the project and payback, if it is achieved, generally comes after many years of operation. This exposes the operation to the vicissitudes of political fortune. Lastly, natural resources often have special emotional and political significance to local populations. For mining companies, any threat to security of tenure and the long-term stability of investment agreements must be considered seriously, and, in fact, this risk represents a "fatal flaw" for many companies and influences decisions about which countries receive investments. The risks relate to different business practices, legal and regulatory systems, as well as the heightened possibility of misunderstandings due to different languages and social customs.

Creeping expropriation, the gradual removal of property rights from a foreign investor through a series of Government initiatives, including new legislation, increases in tax rates, or royalty payments, is a serious risk, as the cumulative effect reduces the economic value of the project to the investor. When the business and operational environment is complex, as it is often the case in the resource sector, there is a greater need for robust governance, as without it there is increased risk of shared service and vendor partnership value leakage.

Possibly as a backlash to globalisation, there is also an increasing nationalistic trend in many countries, often meaning that a foreign Company is viewed unfavourably and operates at a significant disadvantage compared with a local Company. The more successful the business becomes, the more attractive it will be as a target for rent-seeking behaviour. Resource development companies, in particular, are familiar with the situation in which they seem to be regulated far more stringently than their local competitors and receive much more opposition from Non-governmental Organisation (NGOs).

Less well recognised is the role that environmental and social management may play in this scenario. In seeking forfeiture or non-renewal of a mining lease, or leverage in negotiation, agencies may identify environmental issues as the only justification that could withstand the inevitable legal challenge. There have been several recent examples where national or provincial Governments have used dubiously supported breaches of environmental regulations to terminate or expropriate a foreign-owned operation. Typically, accusations are made by local people claiming that their health or well-being has been adversely affected. These complaints bypass the Company's Grievance Resolution system and are made directly to a Government agency that may already be biased against foreign ownership. A cursory investigation follows, often by unqualified investigators, resulting in a guilty verdict without due process. Even where the Company is exonerated, substantial reputational damage may be sustained. Consequently, the owner may be coerced into selling the business to a local Company at well below its true value.

Corruption is a common thread in many examples of attempted and successful forfeiture of assets from foreign companies. This is facilitated in those jurisdictions where companies are required to renew licences on a regular basis; each licence renewal provides the regulator with the opportunity to "hold the Company to ransom." Multinational companies and others with home bases in western democracies have adopted stringent anti-corruption policies. Local companies may have few such constraints. Unfortunately, as this situation continues, more and more successful operations have been transferred from right-minded companies to other less scrupulous operators. Inevitably, the environment and local communities are adversely affected; without foreigners to blame, grievances are less likely to be heard.

Ernst and Young, in discussing the 10 business risks facing mining and metals, identify the LTO as one of the main risks, as stated above. Inherent in Ernst and Young's analysis is that a Company must earn the LTO, which will then protect it from adverse consequences such as loss of tenure. However, the reality is those industrial operations that have done the most in terms of earning an LTO have been singled out for takeover or termination. It is their very success that makes them a target for expropriation. Accordingly, in many situations, insurance against sovereign risks is well worth the cost of premiums. The problem is that policy tenors longer than 7 years can prove to be very expensive from Government and multilateral insurers, and often unavailable on the private market. Capital intensive projects often require 4 or more years to reach profitability and particularly with the significant investments required in earning the LTO.

RISK EVENTS

Identification of the risk events[1] to be evaluated and how these are stated is one of the most critical elements in determining a successful enterprise risk assessment. Commonly, however, an excessive number of risk events are presented. This may result from including risks that are better dealt with at the department level or the inclusion of many generic risk events from various sources with no consideration of materiality. It is acceptable for a small group to filter risk events before assessing risk, typically in a risk assessment workshop. A smaller number of risk events avoid the tendency to "tick and flick" and support a more thorough discussion. There should be no need to include more than 40 or 50 risk events for even complex organisations.

A common error is to indicate a consequence in an event description. This will bias the evaluation of the risk. For example, "light vehicle operation" is acceptable, "serious light vehicle accident" is not. The temptation to "tweak" descriptions year after year should be avoided if at all possible. A different description means a different risk event and comparing the change in risk over time may no longer be possible.

Shown as follows is a sample of risk events used in enterprise risk assessment workshops for a contemporary mining Company. In the inaugural workshop, 127 risk events were assessed. Over 5 successive years, this was progressively reduced to 57 events considered to be material.

- Emissions not meeting Government limits
- Chemical spill in road transport
- Environmental impacts from metals in runoff
- Earthquake
- Plant availability below plan
- Fly rock
- Vehicle accident off-site
- Work-related accidents
- Pandemic
- Premature blasting detonation

1 These are sometimes described as hazards, but this term does not always easily fit non-operational events, such as corruption.

Usually, most time in a risk assessment workshop is spent on evaluating and ranking current risk.

Current Risk is the risk at the time of the risk assessment, with likelihood and consequence being scored based on current risk controls. The current risk generally determines whether a risk is acceptably low or whether additional risk treatment is required. The evaluation of current risk is based on assumptions regarding the reliability and effectiveness of current risk controls.

Inherent Risk is simply the risk of an event without risk controls in place. The value of evaluating inherent risk is that it indicates which risk controls are critical and where management attention should be focused. Inherent risk is sometimes evaluated together with current risk, but in practice, it may seem that almost all enterprise risk events score "Extreme" in the absence of any controls and getting a team to agree on what a "no controls" scenario would comprise can be a frustrating exercise.

Residual Risk is the risk of an event assuming additional or improved risk controls are in place. In theory, residual risk scoring will help identify whether a proposed risk management plan will be effective in reducing risk to acceptable levels. Contrary to common practice, the authors do not support the evaluation of residual risk in risk assessment workshops for the simple reason that enterprise risks are typically dynamic and complex. The effectiveness of a particular risk control will often not be certain until after it has been implemented and can be tested.[2] Evaluation of residual risk is best done *after* additional controls have been implemented, not before. However, where multiple mutually exclusive risk control measures are being considered, some evaluation of residual risks is required to decide which mitigations to fund.

As stated elsewhere, evaluating and ranking current risk is usually the first step in identifying risks that are unacceptable to the organisation and require additional risk treatment. This is a sound approach, but the management team must never lose sight of the risk "iceberg."

---

2 A good example might be a downstream emergency response plan for a tailings facility. Until the capability of site response teams, effectiveness of alarm systems, and local community's ability to speedily evacuate have been tested through drills or simulations, the level of risk reduction provided by such a plan is uncertain.

The Enterprise Risk Iceberg

Current Risks ranked as High or Extreme are usually in clear view of the management team, through the reporting of risk assessment workshop results (they are usually coloured red or orange to attract attention) and a requirement for assigned risk owners to develop and implement risk management plans for each risk, with regular reporting of progress.

Visible Risk

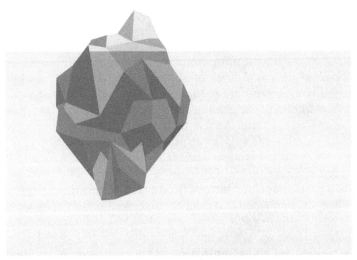

Hidden Risk

Events with high Inherent Risk but assessed as having Low or Moderate Current Risk due to the implementation of a range of risk controls are positioned lower down in the risk register, do not require a risk management plan, and may attract limited attention otherwise. As a result they may tend to disappear from view, but in the event of absent or ineffective controls the real risk may be as high as any other.

RISK SCENARIOS

A given risk event can be described by many different possible scenarios. Confusion often results in a risk assessment workshop when different people have different scenarios in mind when scoring likelihood and consequence. As one example, for a gold mining operation, the risk event "Logistics truck accident" might include a range of plausible scenarios, such as:

Scenario A – Cyanide Truck Collision

- Cyanide pellets carried in an isotainer.
- At worst, collision with a speeding vehicle might result in puncture of the isotainer or damage to a tank fitting.

Release of solid pellets likely to be very limited under most conditions.

Scenario B – Sulphuric Acid Truck Collision

- Concentrated sulphuric acid carried in an isotainer.
- At worst, collision with a speeding vehicle might result in puncture of the isotainer.

Most of the acid could release within a short time.

There might be dozens of other plausible scenarios: collision with a school bus, accident occurring on a busy market day, or collision taking place on a bridge across a small stream. This is the point where risk assessment practice quickly diverges from theory. It is simply not practicable to evaluate all plausible scenarios in a risk assessment workshop.

A simple rule for dealing with multiple scenarios, which seems to work well in most cases, is to identify and evaluate the Realistic (or Credible) Worst-Case scenario. Even though other scenarios will score higher on likelihood (more incidents resulting in more frequent losses), the Realistic Worst Case scenario will almost always be the scenario that delivers the highest risk score.[3]

---

3 As one example, many medical treatment injuries over a period of time will never equal the loss associated with a single fatality. The "credibility" test excludes highly improbable and ridiculous scenarios (e.g. being struck by a meteorite).

Much more complex approaches are available for estimating risk than a risk score based on a single scenario. These may consider a range of scenarios or the uncertainty inherent in estimates of likelihood and consequence. Data required to support these approaches are usually unavailable or incomplete for most enterprise risks, but in any case, there is no need for greater accuracy.

RISK ASSESSMENT

Risk assessment is the method of confronting and expressing uncertainty in predicting the future. Risk is the chance of some degree of damage in some unit of time or the probability or frequency of occurrence of an event with a certain range of adverse consequences. It follows that risk is defined as the combination of the probability (sometimes also termed frequency or likelihood) of occurrence of a defined risk event or hazard and the magnitude of the consequences of the occurrence (i.e., the severity of potential negative impacts). Risk assessment involves the recognition, evaluation, and ranking of risks. For enterprise risks, this is typically implemented through a risk assessment workshop.[4]

The quality of the risk assessment process will determine the success of the entire ERM programme. Errors may result from a range of factors: risk matrices that are poorly designed or do not consider the risk tolerance or exposures specific to the organisation; incomplete listing of risk events to be evaluated; inadequate information to support estimations of likelihood and consequence; human factors such as bias; and so on.

Risk assessments are used to inform decision-making at all levels within companies and by the consultancies that support them. Despite this familiarity, the process of risk assessment can be poorly understood and applied. This perhaps is not surprising when the nature of risk itself is considered. Risk is an abstract concept that is difficult to define and impossible to measure.

---

4 Some companies evaluate opportunities in the same way as enterprise risks, and in the same workshops, reporting opportunities in the same place that risks are reported. For a business with a high-risk capacity, a project with, say, two Class IV risks and three Class IV opportunities might be preferred over a project (of similar financial metrics) with one Class IV risk and no opportunities.

Risk can only be estimated based on predictions regarding future events; a challenge made more difficult because the accuracy of such predictions can only be validated if the risk event were to occur.

Enterprise risk assessments are almost always implemented in workshops involving managers and executives from across the Company, together providing coverage of all relevant areas of knowledge, but ideally limited to eight or so persons. Participation by the Chief Executive Officer (CEO) is essential, and so is appointing an experienced and confident workshop facilitator. This role cannot be filled effectively by someone on the risk assessment team.

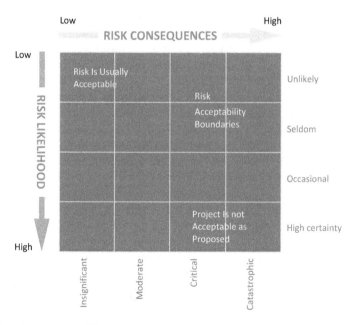

Risk Assessment Diagram

Combining the probability and the severity of those consequences yields an estimation of risk in the form of a simple two-dimensional diagram.

As both probability and severity increase, the risk becomes less and less acceptable. Three concepts taken together describe how a Company treats risk at the enterprise level:

- Risk Capacity is the amount and type of risk an organisation is able to support in pursuit of its business objectives, often expressed in financial terms, such as a maximum amount of debt.
- Risk Appetite is the amount and type of risk an organisation is willing to accept in pursuit of its business objectives.
- Risk Tolerance is the maximum risk that an organisation is willing to take regarding each risk. This may be expressed in quantifiable terms, such as level of invested capital, or qualitatively, such as protection of reputation.

All companies, especially those in the resource industries, need to take risks to obtain rewards. With this understanding, an ERM programme's true purpose is not to mitigate risk but rather the sustainable management of risk in support of business objectives.

To assess risk in practice, three measures are required: (1) rating of risk likelihood; (2) rating of risk consequences; and (3) rating of risk acceptability, all three measures defined according to parameters relevant to a specific risk event. In practice, rating risk likelihood and consequences will produce a risk assessment matrix as illustrated in the figure above, often assigning numerical risk scores (e.g., refer to University of Queensland Enterprise Risk Matrix, 2019, for a more sophisticated risk assessment matrix). For each event, the risk assessment matrix provides a consistent basis for communicating risks and for deciding whether a specific risk is acceptable or whether it is not. Of course, the risk assessment matrix should be interpreted with caution, recognising the oversimplification that it will generally represent.

Both components – probability and consequences – are likely to be at best semi-quantitative, and so each component will to some extent represent judgements based on available knowledge and experience.

Descriptive event probability ratings may range from "unlikely" to "will occur/high certainty." It will be easier to estimate the probability of risk in most risk situations than to judge its consequences. Statistical analysis of rainfall data to determine the design flood is one example. In other cases, estimates can be based on operational experience, e.g., estimating the likelihood of equipment failure.

Descriptive ratings of risk consequences may range from "insignificant" to "catastrophic." Evaluating consequences of risk involves determining the broader implications of a risk event considering all three dimensions of sustainability, that is, considering ecological, social, and economic systems, as well as reputation. The rating of risk consequences will depend on the highest risk in each of these four aspects.

The complicating issue for enterprise risk assessment is the lack of an easily defined measure of what constitutes harmful consequences. In some cases, definitions of damage are intuitive or are laid down in statute (e.g., loss of life or violation of regulatory standards); in others, appropriate criteria need to be based on professional judgements.

Engineering and management efforts are designed to eliminate hazard-related consequences entirely (i.e., moving the probability of occurrence into the category "unlikely"), or if impacts are unavoidable, to move potential environmental risks into the lower-left corner of the risk assessment matrix by reducing the magnitude of the consequences of the occurrence. As such, management measures aim to eliminate hazards, reduce unavoidable consequences to an acceptable level, or both. This begs the question: What is an acceptable environmental risk?

RISK ACCEPTABILITY

The boundaries of risk acceptability are not clearly defined and tailored to the factors influencing the risk's significance. Risks situated in the lower right corner of the risk assessment matrix (risks frequently occurring with catastrophic consequences) are not acceptable. Risks in the upper left corner of the matrix are acceptable, as (1) risk occurrence is unlikely and (2) risk consequences are negligible. The acceptability of risks that fall between these two endpoints depends on a variety of factors identified below.

< Statutory and Policy Requirements – Most jurisdictions have promulgated a wide range of pollution control standards. Where such standards are legally mandated, and a risk assessment demonstrates that an intended activity is likely to breach them, the risk is clearly unacceptable and risk reduction measures are required.

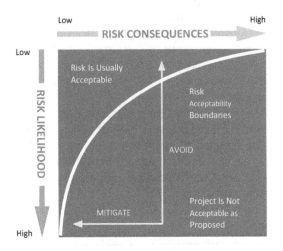

Using the Risk Assessment Diagram to Illustrate
the Aim of Risk and Impact Mitigation Measures

Risk Assessment Diagram

< Loss of Life – Planners, engineers, and operators have a zero
tolerance for loss of life. However, spectacular accidents, such
as Space Shuttle explosion, serve as a reminder that 100% safe
design and operation do not exist even with the best of efforts.

< Value Judgements – Value judgement is subjective. Defining
what constitutes unacceptable harm to an ecosystem, as one
example, is therefore a difficult task and ultimately depends
on what values an individual or society places on ecosystems.
Some may argue that the project site's existing environment
is invaluable and should be maintained at all costs. Others
may hold the opinion that maintenance of ecosystem function is the main objective and that an accidental spill of tailings liquid may not threaten this objective.

< Economic Considerations – Economic factors often have
a significant influence on the acceptability of a given risk.
Economic considerations also extend to considering the provision of costs to mitigate a project impact over and above
those that are provided. Financial consequences due to a delay in project development or disruption of operations are
important considerations when deciding on project risks.

< Social Aspects of Risk – The acceptability of risk can be significantly influenced by a range of psychosocial and political factors. These may include individual risk perceptions and attitudes, cultural values, questions of trust in and credibility of the Company, and questions of equity in risk distribution.

< Reputation Risk – Reputation risks are important, reinforced by events such as the Juukan Cave destruction (BP 01). Reputation risk is of primary concern for large companies that pursue business interests in different parts of the world. Negative publicity in one country can hinder project development in another one. Reputation risk may also influence the capability to raise funds on the financial market.

It takes 20 years to build a reputation and 5 minutes to ruin it.

Warren Buffett

< Residual Risk – Initial investments in risk mitigation can significantly and cost-effectively reduce the vulnerability of a project to identified hazards and can reduce risk consequences. At some point, it may be more cost-effective to transfer the remaining risk or find alternative ways to manage or to finance it than to attempt to mitigate it entirely.

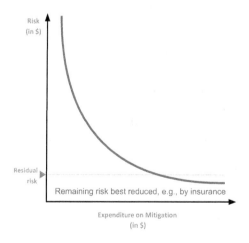

Decreasing Marginal Returns to Investment in Physical Mitigation

USES AND LIMITATIONS OF RISK SCORES

A common misunderstanding is that the risk scores obtained from risk matrices are somehow a measure of risk. This is not the case. Risk scores have no real meaning outside of a risk assessment. Their sole function is to provide a consistent ranking and grouping of risks based on their potential significance to the risk owner. However, in a capital constrained business that is considering more than one investment opportunity, the absolute risk scores for competing projects can be compared. For the comparison to be meaningful, the consequence schemes and approach to assessment must be common to each.

The boundaries in enterprise risk matrices that distinguish between High, Extreme, and other risk classes may be arbitrary and unrelated to the Company's actual risk tolerance. Even if a risk matrix were initially drafted with reference to risk tolerance, this would change over time for any organisation. Enterprise risk matrices are seldom updated to reflect this change.

If the purpose of an enterprise risk assessment is to identify required risk management plans, then the actual risk scores may be unimportant. A management team may decide that the number of risk management plans implemented each year should be limited for effectiveness and resourcing. If, for example, this number is five, then each year, the top five risks are selected for treatment, irrespective of score or class.

As for any tool, risk matrices have limitations (Duijm 2015). These need to be understood and allowed for during an enterprise risk assessment:

- Risk matrices typically comprise colour-coded zones indicating different risk bands, but these zones' boundaries tend to be arbitrary with no relationship to corporate or societal risk tolerance.
- Many risk matrices suffer from range compression. For example, the first four levels of consequence may represent only 25% of the loss in the fifth level. This may lead to similar scores for very different risks.
- Many Company risk matrices include fundamentally different consequence types, and attempt to normalise these across consequence bands. This is a highly subjective process, and some consequences defy comparison, for example loss of life and financial loss.

- To avoid complication, the consequence rating is taken as equal to the highest score from across the consequence classes. After the highest score is recorded, any other scores are ignored. In the example below, both events are rated a consequence score of five, but they are not equivalent risks.

Example 1   Consequence score = 5

Consequence Rating	OHS	ENV	REP	Cost	Compliance
1					
2					√
3					
4			√	√	
5	√	√			

Example 2   Consequence score = 5

Consequence Rating	OHS	ENV	REP	Cost	Compliance
1	√	√			√
2				√	
3					
4					
5	√				

How Consequence is Normally Scored

These limitations highlight the importance of providing time for a review session before a risk workshop ends and the team disperses. Two methods can be applied to rapidly identify results that may involve an error or need further discussion. Firstly, rank the risk scores from high to low and compare each event with its nearest neighbours to check that they appear comparable. Secondly, compare the new score with the score obtained previously for the same event (usually the year before) to highlight any changes that seem contrary to expectations, considering improvements or additions to risk controls in the interim.

It is also common during enterprise risk assessments to encounter events for which there is insufficient available information on which to base credible scoring of likelihood and consequences. In these cases, it is important not to resort to guesswork, but to delay the assessment until the required information is obtained. If crucial information is unavailable, a credible worst-case may be assumed.

A range of human factors will limit the accuracy of any risk assessment. The more dynamic and complex the risks, the more significant these effects will likely be. Some of these factors are discussed in the Appendix in the context of catastrophic risks. Perhaps the most commonplace is the domination of the risk assessment process by a senior or assertive member of the team. The strongly expressed view of a CEO on likelihood or consequence will influence the voting by others around the table. A simple solution to this problem involves applications that support anonymous voting using mobile phones, with scoring collated and presented on a screen in front of the team.

ROLE OF BOARD IN RISK MANAGEMENT

The Board's role has been increasing in importance due to the increasing complexity of our business environment and associated risks. Generally accepted roles of the Board and executives follow:

- Management team led by the CEO bears overall accountability for managing enterprise risk.
- Boards of Directors are not expected to manage enterprise risk. They are expected to provide an oversight role for the risk management systems and processes and continuously review the outcomes of such processes.

- Boards of Directors would usually provide varying degrees of support and advice in support of risk management (e.g., a Director may join an enterprise risk assessment workshop).
- Board and the senior management should be aligned in their understanding of the Company's risk capacity, risk tolerance, and risk appetite.
- Board should take an interest in understanding the Company's enterprise risk capacity. As a minimum, Boards should quantify the Company's debt capacity.

Various arrangements are applied to support a Board's oversight role. In many cases, enterprise risk assessment is delegated to one or more Board special committees.

Environmental liabilities

Risks beyond the balance sheet

Since the 1980s, environmental issues have played a vital role in Merger and Acquisition (M&A) transactions. A rise in costly post-acquisition environmental disputes following the enactment of retroactive site environmental liability laws prompted buyers to evaluate their targets' past environmental performance record before completing transactions; this has become standard practice in most, if not all, industries. The wide range of environmental regulations that apply to industrial operations, coupled with the frequency of regulatory change, also means that new circumstances and thresholds require frequent adaptation of policies, procedures, and activities to avoid unexpected post-acquisition environmental costs.

Most environmental liabilities are manageable if given proper attention and care. The "Polluter Pays" Principle (PPP) that underpins most regulation of pollution affecting land, water, and air provides that the polluter should stomach the costs of environmental liabilities. Indeed, applying PPP means that current (or purchasing) operators or landowners are not responsible for environmental pollution[1] caused by past operators. In practice, however, M&A transactions are often conducted as share sales, whereby all historical environmental liabilities will remain with the Company being purchased.

---

1 Contamination is the presence of a substance that should not be present naturally. Pollution is when the contaminant causes harm to life or infrastructure. The environment can be contaminated without being polluted, but the environment cannot be polluted without it being contaminated.

DOI: 10.1201/9781003134008-12

MERGER
AHEAD

It follows that prospective purchasers need to conduct due diligence to identify the existence of any historical environmental liability that remains with the target Company and its operations. After potential liabilities have been identified and quantified, buyers can determine whether these liabilities will transfer with the purchased business and, if so, may try to obtain an indemnity from the seller. Commonly, the liabilities will reduce the price that a purchaser is prepared to pay.

Some jurisdictions allow transaction parties to contractually agree that the seller will indemnify the buyer for any environmental damage arising due to actions committed by the former operator or landowner. Also, the buyer may be able to claim compensation to recover any loss or liability it suffers due to the existence of pollution. Such contractual arrangements only work if the seller remains in good financial health and business.

In sum, environmental liabilities represent an important issue in negotiations for M&A, particularly in pollution-intensive industries. The bottom line is, the more and the sooner you know, the smarter the agreement you will make.

TAILORING SCOPE OF ENVIRONMENTAL DUE DILIGENCE

No hard and fast rules exist to advise what level of environmental due diligence is appropriate. The buyer decides what level of risk is acceptable. However, well-founded industry practice has developed over the last 40 years or so.[2] Generally, the Phase I Environmental Site Assessment (ESA), or most commonly, the "site audit," involves examination of records (including public databases), interviews with current and past operating personnel, and a site inspection. The Phase II ESA is more intrusive, usually involving subsurface investigation with sampling and laboratory analysis.

The extent and scope of the audit will depend on the nature of the industry, its areal extent or "footprint," its age and history, and the main environmental issues that affect the operations. The form of transaction, and the buyer's objectives, often influence the scope of due diligence, with the regulatory jurisdiction having a vital bearing as well. Primary due diligence concerns include identifying existing contamination, potential clean-up liabilities, and in some cases, estimating remediation costs. Acquisitions of operating businesses or facilities or corporate transactions such as mergers also require evaluation of regulatory compliance and the current and likely future availability of permits to continue and grow the business. Due diligence may estimate capital and operating costs needed to achieve compliance, implement permit conditions, and satisfy other environmental requirements that might include remediation of on-site pollution.

EVALUATING REGULATORY COMPLIANCE IN ACQUIRING ONGOING OPERATIONS

Evaluation of a target's compliance status for environmental regulatory requirements typically includes issues such as whether the business or facility holds all permits and other approvals necessary

---

2 The American Society for Testing and Materials (ASTM) set the industry standard of care for property transfer assessments in the form of Standard E1527-13 (Standard Practice for Environmental Site Assessments: Phase I Environmental Site Assessment Process) and Standard ASTM E1903-19 (Standard Guide for Environmental Site Assessments: Phase II Environmental Site Assessment Process). These are the current versions of the Standards.

to operate and how these authorisations will be transferred as part of the transaction. The buyer will also pay attention to whether the business or facility currently has any significant non-compliance, or history of non-compliance, with regulatory requirements or permit conditions and whether or not any past non-compliances have been reported to relevant authorities in accordance with prevailing regulations.

Generally, regulatory requirements and permit conditions address emissions to air and water (stack and fugitive emissions, wastewater and stormwater discharges) and the relevant control systems, solid and hazardous waste management practices and facilities, emergency planning, and material storage. In addition to identifying regulatory non-compliance issues, the due diligence effort should also attempt to estimate potential costs to reinstate compliance.

WHAT LIES BENEATH – ON-SITE CONTAMINATION

A site may be polluted by one or numerous substances, depending on the nature and source of the contamination. Excavation and treatment of contaminated soil can cost more than US$1,000 per ton, leading to total clean-up costs of many millions of dollars at large sites. Add to this the costs for groundwater clean-up. With these potentially very high costs, existing contamination must be evaluated before a transaction can be completed. This commonly takes the form of a Phase I ESA carried out by a consultant. Should this study identify the presence of significant contamination, a Phase 2 assessment follows, enabling the extent and severity of contamination to be evaluated.

On-site contamination is industry-specific. Historically, exploration for and production of oil and gas, for example have caused soil and water pollution through improper disposal of produced saline water, oily sludge, and drilling mud. Accidental release of hydrocarbons and abandoned oil wells that were orphaned or not correctly plugged are additional pollution sources. Most old oil fields will have some combination (and perhaps all) of these issues.

It has also been well established that naturally occurring radioactive material (NORM) may accumulate along the oil and gas production process chain. Scale in flow lines may contain NORM,

predominantly compounds of radium. Components such as well-heads, separation vessels, pumps, and other processing equipment can also become contaminated with NORM, accumulating in the form of sludge or scale. Removal of pipes and components exposes workers to radiation, especially when they clean parts for reuse.

Decommissioning of oil wells, processing and storage facilities, offshore platforms, pumping stations, and flow lines escalates potential environmental liability costs. Decommissioning liabilities can be large and readily observable, but somehow many operators would rather ignore or deny them than face these realities.

Environmental liabilities related to mining of nonfuel minerals are likewise numerous and diverse. Acid mine drainage is notorious for its potential impact on the environment that may not be noticed until after active mining operations cease. Effects can persist for decades and even centuries. How to prevent – in perpetuity – the release of toxic contaminants (mainly metals) from various mine facilities (such as abandoned open pits, waste rock dumps, mine adits, and tailings impoundments)? And how to set aside funds to ensure that the costs of reclamation and closure can be covered?

OFFSITE CONTAMINATION ON FORMER SITES

The PPP is logical and straightforward; those who cause pollution and environmental damage should be responsible for its remediation. PPP also suggests that protecting against clean-up liability on occasion may require looking beyond the sites that are part of the M&A transaction. Under PPP, a former owner or operator of a property may be held liable for pollution at its previous locations if the disposal or release giving rise to the pollution occurred during the period in which that entity owned or operated the property. To the extent possible, buyers should evaluate the existence and the nature of operations at the target's former operational sites. This may require research of a Company's history extending back through several changes of ownership and control.

OFFSITE DISPOSAL OF HAZARDOUS WASTES

The "Cradle to Grave" concept was first introduced in the USA through the Resource Conservation and Recovery Act of 1976 (RCRA) and reinforced by the Comprehensive Environmental Response, Compensation, and Liability Act of 1980 (CERCLA – the Superfund Law). Adopted in some form by many jurisdictions in the decades that followed, this regulatory innovation designates a hazardous waste generator as responsible for its waste from generation through ultimate disposal and beyond, including offsite migration of contaminants. For environmental offences in the USA, this overrides the applicability of limited liability.

Nothing relieves a generator of this responsibility, which is retroactive and assigns liability for actions that were legal and common industry practice at the time. Bankruptcy is not a shield – share ownership is traced to living individuals, heirs, and affiliated companies. There is no expiration date or time limit, and hiring someone else to transport and dispose of your wastes does not transfer responsibility. In-house mishandling and casual disposal of production wastes led to the concept of site liabilities and Phase I and II ESAs – as has been said, sometimes waste falls out of the cradle (hence RCRA). Problematic commercial waste disposal operations left severe residual problems for the Government. In response, Governments have assigned liabilities to companies associated with formerly unregulated practices of which companies may have been unaware, at sites they never saw – so sometimes waste comes back from the grave (hence CERCLA). "Cradle to Grave" indeed.

The bottom line is, if you generate hazardous waste, you can and will be held responsible for its improper transportation and disposal. And a waste can be judged "hazardous" simply because there is a lot of it. Environmental clean-up liability is often "jointly and severally" imposed without regard to the degree of fault. It makes no difference that the generator had no part in the actual placement of hazardous waste at the disposal site or in choosing the location to which the waste was shipped. Prospective buyers should critically review the target's (and any publicly available) documents and records concerning storage, transportation, treatment, and disposal of its hazardous waste, extending well back in time. Weak or venal Governments may delay accountability, but contamination problems seldom diminish or disappear on their own.

IN-MIGRATING CONTAMINATION

In most jurisdictions, you are not held responsible for pollution that originates offsite and migrates onto your site (unless, of course, your operations have contributed to the pollution). However, simply because there is no obligation to remediate does not mean that migrating contaminants can be ignored in an M&A transaction. The presence of contaminants may require special protection for construction workers during future earthworks; excavated soil or pumped groundwater could incur high treatment or disposal costs. Even if no earthworks will occur, migrating contamination in the form of vapour may create a health risk to employees that could require costly mitigation. Moreover, migrating contamination may affect down-gradient landowners and residents. Even though the actual source of pollution is offsite, people adjacent to your site are likely to address their concerns to you.

BUILDING MATERIALS AND INDOOR AIR QUALITY

Building and decorating materials are additional sources of potential liabilities and indoor pollution. Asbestos-containing building materials are prime examples. Others include natural radioactivity in the form of radon gas from building materials and underlying soils; and volatile toxic chemicals in the air released from interior decorating materials, such as textile carpeting, paints, dyes, and glues. The following brief and non-exhaustive discussion illustrates the range of potential environmental liabilities from these materials.

Although asbestos was discontinued in building products in most jurisdictions from the late 1970s, investigations of asbestos liabilities remain an essential part of the due diligence process in acquisitions, as many structures predate these changes. M&A actions affecting businesses or assets with potential latent asbestos liabilities contain pitfalls for two main reasons.

First, businesses that used asbestos in the past are at risk of occupational health claims from employees, including retirees. The maximum possible future exposure can be estimated by taking into account the period in the Company's history when asbestos was used in its manufacturing processes and the typical employment history of the class of workers who may have been exposed. This problem has been "aging out," as most workers who began their careers in the 1970s are retiring, but it persists in parts of the world where the regulatory changes were delayed.

Second, asbestos was historically used in products and places not readily apparent, and many older buildings still have asbestos-containing materials (ACM). ACM were used in various building applications such as roofing, ceilings, insulation, and floor tiles, due to the many beneficial properties of asbestos, including fire resistance, chemical resistance, and fibre strength. ACM also remain in many industrial boilers, furnaces, and kilns that have long service lives. Recognised asbestos abatement options include managing in place, encapsulation, enclosure, and removal. Asbestos abatement costs vary but are often material post-acquisition environmental costs.

The mere presence of asbestos in a building is usually not a transaction deal-breaker, as long as there is a full understanding of the location and condition of any ACM, and potential latent asbestos liabilities related to past and current building users.

Radon, a colourless, odourless, and carcinogenic gas, comes from the natural radioactive breakdown of uranium. Granite may release relatively high quantities of radon, whereas most sedimentary rocks have far lower concentrations. Radon can be present in soils and groundwater, as well as building materials. Radon can

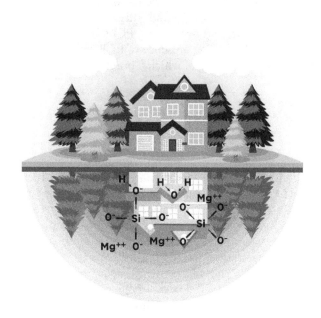

infiltrate indoor air through basement floors and walls, or water used in the building. Testing air at the building's lowest level will identify potential radon risks. Costs for implementing mitigation measures (e.g., vent pipes and fans to prevent radon from seeping into the building) are generally not material in the context of most transactions.

Some European countries banned lead-based paint in buildings in 1909 when scientists highlighted the significant health risk of lead for young children and pregnant women. However, as with asbestos, lead paints were not widely discontinued until the mid- and late 1970s. Many residential and commercial structures, mainly those built before 1950, may still contain lead-based paint. Children and pregnant women are unlikely to be prominent occupants of industrial buildings. Nevertheless, potential buyers need to be aware that a failure to identify or sufficiently abate lead-based paint hazards ultimately can result in personal injury claims. These claims may assert lead poisoning based on interpretations of negligence, strict liability, or breach of express or implied warranty.

Polychlorinated biphenyls (PCBs) were also widely used in the past in electrical transformers and still exist in some older industrial sites. Manufacturers stopped using PCBs in most of the world by the mid-1980s. Again, the presence of PCB is unlikely to be a deal-breaker but is something of which a purchaser needs to be aware.

EMERGING ISSUES IN ENVIRONMENTAL DUE DILIGENCE

In jurisdictions with robust enforcement and more mature environmental legislation (particularly in the USA, the European Union (EU), Canada, Australia, and Japan), Contaminants of Emerging Concern (CECs) are a developing issue in environmental due diligence and ESAs. CECs include chemicals and compounds that, while not necessarily new, are newly identified as posing perceived potential, or real, threats to human health and the environment. There have always been, and will likely always be, CECs to be considered in due diligence. The list is ever-evolving. Many compounds that are intensely regulated now for their harmful nature were widely and casually used and entered the environment in the quite recent past. Asbestos, dichlorodiphenyltrichloroethane (DDT), and PCBs were clear examples two generations ago, though now infrequently encountered. Today, endocrine-disrupting chemicals (EDCs) such as phthalates, parabens, and triclosan are CECs.

Pollution should never be the price of prosperity.

Al Gore

Shifting goalposts of environmental compliance are another emerging issue in environmental due diligence. In Germany, as one example, a recent amendment to the Law on Recycling and Waste (now the Recycling Act) has forced German companies to revisit their environmental compliance programmes. Producers are now obliged to reduce the overall effects of resource use and improve the efficiency of use, in particular by recycling and avoiding waste generation. The Industrial Emissions Directive (IED) in the EU is a second example. IED requires companies to adjust their operations to new emission thresholds and implement the best available techniques to prevent and control industrial pollution.

Another environmental issue fast becoming a concern in environmental due diligence is climate change (BP 13). Generally, climate change risks fall into two categories: regulatory and physical risks. Regulatory risks relate to the ambitious plans for reducing GHG emissions by numerous Governments and other organisations. Considering the various GHG reduction initiatives, the brunt of GHG emission reductions will fall on business owners and operators (e.g., in one form of carbon tax). Business activities and assets are also at physical risk from climate change.

Environmental due diligence should identify the operations and infrastructure of the target organisation sensitive to climate change and decide based on mitigation costs and threats to reputation as to whether these are significant relative to other sources of risk.

Finally, institutional investors, including many private equity funds, increasingly incorporate environmental, social, and corporate governance (ESG) considerations in their respective due diligence processes to better manage risk and create value (BP 04). For example, the United Nations-supported Principles for Responsible Investment ("UNPRI") Initiative has seen a remarkable increase in the number of signatory parties over the past 5 years. As of 2020, it represented approximately 1,200 asset owners, investment managers, and professional service partners, with nearly US$35 trillion in assets under management.

ALLOCATING ENVIRONMENTAL RISKS IN TRANSACTION AGREEMENTS

As the breadth of environmental liabilities has become more apparent, contracting parties attempt to apportion risks of these liabilities between themselves. Risk allocation issues arise not only in transactions between buyers and sellers but also between lenders and borrowers. Now and then, the resulting allocation of risk is a function of the type of transaction or the parties' relative bargaining positions. At other times parties genuinely attempt to divide responsibility for perceived risks in a manner considered fair for both sides.

Contractual provisions that allocate environmental risks can take various forms, including indemnities, hold harmless clauses, exculpations, disclaimers, "as is" clauses, survival provisions, and releases. An environmental indemnity clause could read as follows: "Seller shall indemnify Buyer against, and hold Buyer harmless from, any damages, costs or claims, incurred by, or which may be made against, Buyer in connection with any pre-existing pollution."

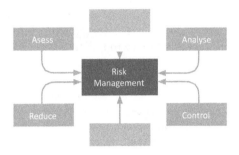

ENVIRONMENTAL INSURANCE POLICIES

Environmental insurance policies serve to fill the gap in coverage for environmental liabilities in general public liability policies. Insurers remain cautious about providing long-term (10 years) coverage for M&A transactions; policies covering up to 3–5 years for new and pre-existing pollution conditions are standard. Along with shorter policy terms, insurers seldom offer limits exceeding US$25 million, requiring more recent environmental due diligence information and restricting coverage on properties slated for development or redevelopment activities.

A further type of policy that may be relevant in certain circumstances provides cover for pre-existing contamination, such as contamination present at the insured site but unknown to the insured when the policy incepts. Alternatively, the contamination is known to the insured, but the insurer is prepared to take the risk that remediation will not be necessary during the policy period.

ENVIRONMENTAL RISKS/LIABILITIES REPORTING REQUIREMENTS – DAMNED IF DO AND DAMNED IF DON'T

Environmental reporting is governed by law. In M&A transactions, business leaders may encounter problems determining which information unearthed during due diligence must be reported and which can legally and responsibly be omitted from required reporting to Government authorities. Most jurisdictions have provisions concerning reporting requirements for identified soil and groundwater contamination. The bottom line is that if an environmental due diligence (or any other activity during operation, e.g., construction earthworks) suggests the presence of site pollution, there is likely to be a legal duty to report the contamination promptly and fully. Failure to do so may be an offence incurring a fine or leading to prosecution.

END OF LEASE/TENEMENT CLEAN-UP

The need to identify and address environmental legacy issues is not limited to M&A transactions or reported pollution. In the extractive industry, at some point, operators or contractors will relinquish the lease area (also termed tenement or contract of work) to the issuing authority or another party. Lease relinquishment can occur when the economically mineable resource has been exhausted, or it happens by default when the lease agreement expires. Most lease agreements require the operator to return the lease area free of environmental pollution and long-term environmental liabilities. This requirement can lead to significant closure operations, sometimes for extended periods as is discussed in more detail in BP 15.

There is the need to obtain sign-off by all concerned Government authorities. The process will typically involve a final evaluation of the site to ensure it has met all the designated performance and outcome criteria. The Closeout Audit may require a third-party assessor or a panel of experts/stakeholders who can undertake the final review and provide a recommendation to the regulatory authorities. The aim is to ensure the site does not leave long-term environmental or social liabilities for the Government, as lease termination ultimately enables the operator to relinquish responsibility for site management.

Strategies for climate change risk management

Stemming the tide

For the past four decades, climate scientists have warned that the world faces serious, even catastrophic, consequences as a result of climate change. Much of the world has warmed significantly over this period, and this warming has been attributed in large part to increased emissions of greenhouse gases (GHG) – notably carbon dioxide – as a result of fossil fuel usage for power generation and transportation. In 1998, the Intergovernmental Panel on Climate Change (IPCC) was established by the United Nations to provide policymakers with regular scientific assessments on climate change, its implications and potential future risks, as well as to put forward mitigation and adaptation options.

The IPCC has produced a range of predictions obtained from numerical models incorporating atmospheric and oceanographic processes and increased carbon dioxide concentrations based on various fossil fuel usage scenarios. The worst of these scenarios, with ever-increasing fossil fuel usage, predicted catastrophic outcomes due to changes in weather patterns and rapidly rising sea levels. In response to these predictions, in 2015, 197 countries signed the Paris Agreement, a legally binding international treaty on climate change. Even conservative Governments that initially resisted such moves have now adopted challenging schedules to attain so-called carbon neutrality.

Business leaders can ill-afford to ignore climate change and its potential consequences. Besides, we must explain climate change risks and adaptation moves, as shareholders increasingly demand clear insight into how companies are managing climate change risks, whether real or perceived. We business leaders have no choice but to ensure our companies account for climate change impact

DOI: 10.1201/9781003134008-13

risks, adopt suitable management responses, and communicate climate change adaptation strategies well.

> We are the first generation to feel the sting of climate change, and we are the last generation that can do something about it
>
> Jay Inslee

Climate projections are uncertain. Though there may be partisan debate around the level of climate change, climate change is real and happening rapidly. We can debate the theoretical, but actual business leaders are action-oriented and focus on what can be changed now, not in the future, to make their companies more sustainable and "do good" along the way.

5 coolest years	5 warmest years
1885, 1888, 1892, 1919, 1963	2006, 2007, 2011, 2014, 2017

Our warming world in stripe form – UK Annual Temperature 1884 to 2020 (Source: 'warming stripe' graphics produced by Professor Ed Hawkins (University of Reading), Royal Meteorological Society)

Some business leaders may argue that their business efforts are negligible and undercut by the lack of clear political and economic signals to decarbonise the economy. This story places the responsibility solely on the public sector, as it suggests that significant change at scale cannot happen without political intervention. It ignores that the private sector is responsible for the majority of GDP and job creation. The private sector must take ownership of its impact on the environment while working with the public sector to accomplish change.

CLIMATE CHANGE RISKS TO INDUSTRY SECTORS

It is fair to say that business leaders for many years resisted calls to respond to climate change, buying time to avoid major expenditure; this approach largely worked until recently. However, the global shift toward greater environmental awareness has now reversed this situation. At the strategic level, business leaders are well-advised to review the characteristics of their Company's business sector, to take into account likely changes in future market and regulatory conditions due to climate change, and to adjust investment patterns. At the same time, investors need to review portfolio composition to accommodate regulatory changes and possible changes in resource supply and demand.

Mining of coal and its use in power generation are an obvious case in point – the main risk from climate change to the coal industry will likely come from growing concerns of society that both increase the difficulty in obtaining approvals for new coal mining projects and coal-fired power plants and regulatory changes and subsidies that favour less carbon-intensive energy sources. Work to develop clean coal technologies may somewhat ameliorate this risk; however, the processes of coal mining and coal combustion will likely face increasing societal pressure, while ethical investing tendencies increasingly favour renewable energy sources.

Societal pressure is affecting companies' ability to raise project financing, as lenders distance themselves from financing coal mining and coal-fired power plants. Carbon tax – a fee imposed on burning carbon-based fuels (coal, oil, gas), and as such the principal policy for reducing and ultimately eliminating the use of fossil fuels – is likewise a response to societal pressure. Policies addressing climate change and GHG emissions vary widely around the globe, both in substance and in their pace of implementation. Business leaders need to actively monitor emerging GHG/climate regulations to make an informed transition to non-carbon fuels and higher energy efficiency.

CLIMATE CHANGE OPPORTUNITIES FOR INDUSTRY SECTORS

As business leaders identify climate change risks, they equally need to spot business opportunities. Change drivers – the affordability of renewable energy and Government policies to curb GHG emissions – have created an environment that favours low-carbon technologies with associated business opportunities. The demand for minerals and metals used in low-carbon technologies will only gather pace in the future, as will sustainable and reliable extraction and production operations. This example illustrates that extractive industry sectors and companies are inextricably linked to both causes of and solutions to the climate crisis. Mining and mineral processing contribute to GHG emissions. Still, a low-carbon economy requires technologies that drive increased demand for specific metals – copper, lithium, zinc, lead, rare earths, cobalt, manganese, and nickel, while the needs for aluminium, copper, and iron are not going to diminish. Similarly, with hydrocarbons, whatever reductions occur in the use of oil and gas to generate power and as transport fuels, the use of hydrocarbons as feedstock for plastics and petrochemicals is likely to continue unabated.

Society does not always acknowledge the mining industry's contribution to the reduction of GHG and other gaseous emissions (Spitz and Trudinger 2019). Mining produces lightweight metals (aluminium and magnesium) that have enabled manufacturers to produce far more fuel-efficient vehicles than their predecessors. Similarly, mining provides metals such as platinum and palladium, used in catalytic converters that improve combustion efficiency and reduce potentially toxic emissions. Furthermore, mining produces the materials used in the manufacture of computer chips and electronic fuel injection systems, which have improved vehicle fuel efficiency. Mining also provides the metals used in grid-scale battery installations, making uneven power sources such as wind and solar capable of offering reliable base-load electricity.

Furthermore, it is also not widely acknowledged that clean energy technologies such as wind farms, solar panels, and batteries are more material intensive than current traditional fossil-based energy systems. Recycling of minerals will not meet this growing demand, and the world will continue to rely on the mining of minerals and metals.

ENERGY AND GHG EMISSIONS MANAGEMENT

Business leaders need to acknowledge if their Company is energy-intensive and a significant emitter of GHG, as is true for most mining operations. However, this depends on the type of resource mined and mine design. In an age of global environmental awareness, business leaders must demonstrate that improving energy efficiency and reducing GHG emissions are priorities for their companies to limit environmental impacts and reduce operational costs.

Investment in energy efficiency is a straightforward way, and the low hanging fruit to cut carbon emissions. The technology we need to make a change exists today – e.g., cleaner, more efficient vehicles; higher minimum efficiency standards for lighting; and upgrading the efficiency of older buildings through improved insulation.

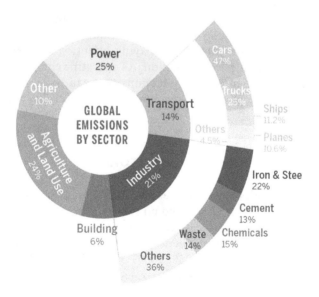

Source: Emissions data is from the IPCC's Fifth Assessment Report. Working Group III, 2014, and refers to shares of total global GHG emissions. The split between cars and trucks in road transport emissions is based on the International Energy Agency's (IEA) Energy Technology Perspectives 2017, since this is not given in the IPCC source.

A small number of indicators will confirm that a Company has established a comprehensive system for recording and managing energy use and GHG emissions (Mining Association of Canada). A Company should be able to demonstrate its management system including assigning accountability to senior management. The Company should review energy data regularly, and energy efficiency should be well-integrated with operator actions. Furthermore, a Company should arrange for energy awareness training and track and report energy use and GHG emissions data, both internally and externally. Finally, in the interest of continuous improvement, the Company should institute and meet performance targets for energy use and GHG emissions.

Unfortunately, the nature of most mining operations means that energy usage per unit of production increases over the life of the project. This is due to the sequencing of activities, whereby the closest/shallowest ore is mined early in the project life, with increasingly deeper, more distant minerals mined as the project proceeds. This applies to strip mines, open pit mines, and underground mines. Anti-mining critics pay close attention to energy usage as presented in Company Sustainability Reports. Year-to-year increases in energy consumption are cited as evidence that a Company is not paying attention to energy management and emissions reduction. As operating companies know, there is a limit to energy reductions that can be achieved, and, in the absence of offsets, reaching carbon neutrality may prove impossible.

CLIMATE VARIABILITY IN FUTURE

Most companies have practices and strategies in place to deal with routine climate variability and no sectors more so than the extractive industry, which by its very nature is affected by geological, hydrological, and atmospheric phenomena. In managing climate variability in the future, companies obviously cannot assume that the prevailing climate will more or less resemble its characteristics over the past 50 years. Climate change is invalidating such simplistic ideas, with both changes in average meteorological conditions and altered spatial and temporal distribution of specific parameters (notably rainfall) even when annual averages remain mostly unchanged. Such climate changes arguably are more relevant to projects with a long operating life compared to those with shorter investment horizons.

To judge if climate change is a significant risk, business leaders need to predict future climate variability for the geographical area of their operations. They can access widely and publicly available long-term projections of climate change over the next few decades. However, predictions of future climate conditions have proved unreliable; there are so many variables involved that it is impossible to make accurate climate predictions for longer time-frames. This does not mean that business leaders may avoid their responsibility to understand climate change and plan responses to potential future scenarios.

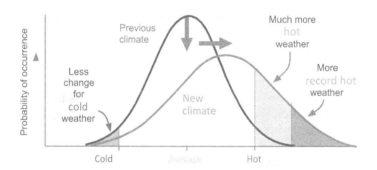

CLIMATE CHANGE RISKS TO OPERATIONS

Some business leaders may argue that climate-related risks are systemic and too complex to assess at the business or industry level. Those leaders may not recognise that climate change will create winners and losers. Those with a negative mindset will suffer. At the same time, those that take proactive steps to assess climate-related risks will minimise business costs from disruptions or interruptions due to extreme weather. Errors and mistakes are inevitable, but the biggest mistake is to ignore the challenge.

At the operational level, business leaders should identify their activities and assets at risk from climate change. To do so, they need to spot (based on their professional knowledge) operations and infrastructure of the organisation that are sensitive to climate change, and judge whether this is a more significant source of risk relative to other sources. Assessment of possible impacts to assets includes those resulting from likely increased frequency or severity of storms, rainfall-induced landslips, rising sea levels, temperature increases (or decreases), and reductions or interruptions in water supplies.

Increasingly it is clear that rainfall patterns are changing with the climate. While rainfall totals may not change significantly, the trend appears, particularly in the humid tropics, to point in the direction of wetter rainy seasons and drier dry seasons. Probably, more importantly, there is ample reason to suspect rainfall event intensities in many areas are increasing while they decrease in others. Past design of road grades and drainage (pavement drainage, culverts, box culverts, and bridges) and erosion and sediment controls (size and number of sediment traps, overflow structures, and diversion ditches) may not cope with future rainfall concentrations and runoff volumes. Ineffective erosion and sedimentation controls can have consequent impacts on downstream water users. In addition, the inundation of low-lying and coastal areas can cause long-term damage to project infrastructure and facilities, requiring expensive reconstruction or closure.

Just as there will be too much water in some areas, there will be water shortages in others, and the need for sustainable water management within most industry sectors is becoming even more critical. So what can the mining industry, an example of a major water consumer, do to conserve this precious resource?

Note that encouragement shouldn't be needed, as for miners a lack of water can mean an operational shutdown. Mining uses water primarily for mineral processing, dust suppression, slurry transport, and employee needs. Mining operations source water from groundwater, streams, rivers, lakes, or commercial water service suppliers. But mine sites are often located where water is already scarce, and, understandably, local communities and authorities often oppose mines' heavy usage of water from these sources.

As the figure shows, risk is the intersection of hazards, vulnerability, and exposure. While business leaders cannot control or

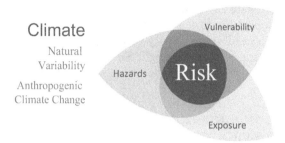

Climate

Natural
Variability

Anthropogenic
Climate Change

Vulnerability

Hazards   Risk

Exposure

reduce the hazards, they can maintain awareness and ensure regular flows of relevant information on their nature and magnitude. Planning, engineering, and management decisions can reduce vulnerability to some extent if hazards are appreciated. Controlling exposure may not be possible in the near time but can affect long-range siting, investment, and technology decisions.

COMPANY IMPACTS ON HOST COMMUNITIES AND CLIMATE CHANGE

The extractive industry has significant impacts on water resources, by depleting water supplies and polluting them with discharges or seepage from tailings and waste rock impoundments, and disturbed mining areas. The issues are discussed at length in BP 09; global sustainability awareness and environmental regulations that have encouraged greener water practices in the industry are now considered essential to public acceptance of modern mining. One way of "greening" mining operations is by capturing or diverting surface water streams and runoff that may wash contaminants into the environment. Other ways to prevent seepage and stop the escape of harmful substances are to cap piles of potentially acid-generating waste rock and to install liners under placement and stockpile areas. Water recycling is commonly applied to reduce the total volumes of water that must be withdrawn and discharged.

Concerns for climate change will only increase pressure on business leaders to commit to better water stewardship. Perceptions of overuse or polluting practices will inevitably generate opposition to a project. It follows that business leaders should routinely implement and communicate water management strategies to mitigate environmental impacts, from project development through operation to closure and restoration. Business leaders may consider exceeding regulatory compliance requirements to maintain their status as good neighbours to host communities and to provide access to water for these communities, if necessary.

HOW COMPANIES COMMIT TO CLIMATE-PROOFING

Business leaders' response to questions of how their Company prepares for climate change adaptation will boil down to assuring stakeholders that climate change adaption, or climate-proofing, is integral to Company planning and operation:

> First, our enterprise risk management process provides corporate oversight for identifying significant risks to the Company and ensuring mitigation plans are in place.
> The method includes an annual risk review with executive management and the Board of Directors that identifies risks inherent in the business, including climate change risks.
> There is a clear understanding of how market and regulatory conditions due to climate change will impact possible demand and supply, positively (opportunities) and negatively (threats).
> Second, improving energy efficiency and reducing greenhouse gas emissions are permanent priorities for our Company.
> Third, our Company has adjusted existing practices and strategies to deal with routine climate variability, including climate variability in the future. Climate change is part of our disaster and environmental risk screening, a proactive approach to considering short- and long-term climate risks in the operational planning process.
> Fourth, our Company has embraced numerous options for climate change adaption, all of which are well characterised (i.e., have known strengths and weaknesses).
> Finally, and most importantly, our Company understands its impacts on host communities, and how these impacts may change due to climate variability in the future. Our Company has the commitment and the resources to ameliorate these risks to the extent necessary.

In sum, we believe our Company's current risk management and business planning processes are sufficient to mitigate the risks associated with climate change. These processes are appropriate for monitoring and adjusting accordingly as climate policy developments and impacts unfold.

Crisis management

Don't wait for a crisis to come up with a crisis plan

While the common assumption is that crises are rare events, the reality, within the typical planning horizons of companies, is that crises are commonplace. Extractive industry experience indicates companies with more than 5,000 employees on average are likely to experience a crisis every year (PwC 2019). And unfortunately, no two crises are the same.

Companies that are poorly prepared to manage these unfortunate events will be at risk of significant impacts up to and including closure. Companies that can deal effectively with crises will reap a range of benefits, including lower operating costs and enhanced social licence to operate.

All companies seeking resilience in the face of crises should have an effective crisis management system based on industry-leading practices. Somewhat surprisingly, there are few sources of guidance available for implementing crisis management systems at the Company level and a virtual absence of benchmarking of crisis management practices across the extractive industries.

The term "crisis" has a wide range of meanings, is imprecise, and is overused. Often crises are described as unfortunate events impacting an organisation, but it is easy to identify examples of crises involving potential rather than actual impacts. For example, the development of unsafe conditions within a tailings facility embankment would be seen as a crisis by most mining companies. This is an essential understanding, as organisations that are quick to recognise the early indicators of a developing crisis are likely to be much better positioned to implement an effective crisis management response. Sometimes this early period is referred to as a pre-crisis stage. Actions implemented in this period are critical in limiting impacts and should always be part of a crisis management response.

DOI: 10.1201/9781003134008-14

ISSUE AND CRISIS MANAGEMENT RELATIONAL MODEL

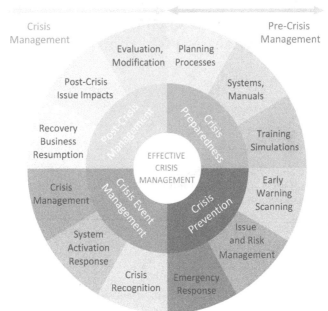

(Based on Jaques 2007).

CRISIS VERSUS NORMAL OPERATIONS

Some important factors that differentiate crisis management from routine operational management need to be considered when developing a Crisis Management Plan (CMP) and establishing a Crisis Management Team (CMT):

- Command and control structure that is leadership-based and directive rather than consultative is required.
- Decision-making and task execution cycles are usually much quicker.
- Available information is often incomplete or inaccurate.
- Assigned responsibilities and tasks may be very different from those faced in normal work.
- Consequences of mismanagement can be severe.

- Impacts on stakeholders are almost always involved.
- Media exposure is likely, and this may include heavy criticism of the Company.
- Company executives may be exposed to criminal prosecution about decisions made.
- Crisis management is often stressful and fatiguing.

Hence, it is hardly surprising that many companies perform poorly in managing crisis events. More surprising is the complacency or overconfidence that many companies display in preparing for crises.

Crisis can be categorised in several ways, most commonly by the nature of the event and its likely impacts, as shown below.

Operational	An incident or series of events resulting in a significant disruption to normal operations. Examples include breakdowns, structural failures, fires, supply chain disruptions, illegal blockades, pollution of the environment, and natural disasters.
Financial	Significant falls in profitability or asset value, threatening cash flow, profitability, share price, and ultimately business closure or receivership.
Personnel	Incidents involving unethical or illegal conduct by an employee or Director. Examples include fraud, corruption, malfeasance, and sexual harassment.
Organisational	Situations where a Company has significantly wronged its consumers, employees, or the public. Exposure of this action may incur loss of sovereign support, cancellation of exploitation rights, expropriation, reputational damage, shareholder activism, loss of market share, fines, and criminal prosecution.
Information Technology	System failure, data loss, cybercrime.

In the case of a crisis involving an operational incident, it is almost always the case that a chain of causation and multiple contributing factors were involved, causing multidimensional impacts.

In 2019, Pricewaterhouse Coopers (PwC) conducted a crisis management survey of companies from 25 industries and 43 countries, receiving 2,000 responses from senior management and executives. Nearly two-thirds of respondents had experienced at least one corporate crisis in the 5 years preceding the survey. The average number was three. Over one-half of all respondents experienced operational crises; one-third experienced information technology crises. Organisations with more than 5,000 employees most commonly encountered cybercrime, natural disasters, leadership issues, and misconduct crises.

The key learnings from this data for any Company are that business crises are inevitable, are not uncommon, involve any aspect of the business, and need to be addressed by proper planning and resourcing before the event.

CRISIS MANAGEMENT MODELS

A crisis management model is a conceptual framework to describe how crises develop and how they can be managed. There is a wide diversity of such models, and there are several relevant international standards, including BS 11200:2014 Crisis Management Guidance and Good Practice and NFPA 1600 Continuity, Emergency, and Crisis Management. Several models that usefully illustrate some critical aspects of crisis management at a Company level follow.

SCENARIO-BASED VERSUS CAPACITY-BASED MODEL

Scenario-based crisis preparedness is based on the preparation of detailed response plans addressing a range of crisis events. When a crisis occurs, the appropriate response plan is selected and followed, much like a Standard Operating Procedure. Increasing rates of social change, technological development, and operational complexity have led to increasingly complex and unpredictable crisis events for which scenario-based planning is minimally effective.

A well-known aphorism attributed to Field Marshal Bernhard von Moltke is that no plan survives first contact with the enemy.

He also wrote the following commentary that is as applicable to crisis management as it is to warfare:

> Certainly the commander in chief will keep his great objective continuously in mind, undisturbed by the vicissitudes of events. But the path on which he hopes to reach it can never be firmly established in advance. Throughout the campaign he must make a series of decisions on the basis of situations that cannot be foreseen... Everything depends on penetrating the uncertainty of veiled situations to evaluate the facts, to clarify the unknown, to make decisions rapidly, and then to carry them out with strength and constancy.

A more robust and flexible approach to crisis preparedness is based on crisis management teams' capacity-building through measures such as team selection criteria, role-based checklists, competency-based training, response drills, crisis simulations, and the use of crisis management software. Procedures are still necessary, but more as references in preparing for crises rather than for use during events. There is obviously an essential place within crisis management for emergency response Standard Operating Procedures addressing technical areas such as spill response and medical evacuations.

PROACTIVE VERSUS REACTIVE CRISIS MANAGEMENT MODEL

The different levels of crisis management maturity in organisations can be described by a proactive versus reactive crisis management model comprising the following stages, from most to least advanced:

Pre-emptive	Planning and preparation are in place, including a formal crisis management system and a trained CMT. Beyond this, however, the organisation has also established effective enterprise risk and issue management processes to minimise the occurrence of crises and their impacts.

Proactive	Planning and preparation are in place, including a formal crisis management system and a trained CMT. These are activated and effective early in crisis events. Initiatives are taken early in a crisis to shape how events unfold.
Responsive	Crises may be unexpected. However, the management team's rapid response, including effective situation assessments and implementation of response plans, effectively deals with most crises.
Reactive	The response to crises is defensive and reactive – often panic-driven or poorly organised with limited objective thinking, systematic problem-solving, and planning.

ISSUE AND CRISIS MANAGEMENT RELATIONAL MODEL

An issue is any unplanned development or situation that may result in losses for an organisation if not resolved through management intervention. Under the issue and crisis management model, crises result from inadequate management of issues or enterprise risk.

Resource companies generally operate in high-risk environments. Management teams that are not successful in managing enterprise risk or operational issues can be expected to be crisis-prone. However, it is also true that risk can never be eliminated; some enterprise risks are unmanageable or unknown, and many crises faced by companies do not result from unresolved issues.

INCIDENT COMMAND SYSTEM MODEL

The incident command system model is a management framework that entails organising crisis management outcomes into functional groups addressed by defined team roles and accountabilities. In the event of a crisis, members of the management team delegate their normal duties and assume a role in the CMT. Each Company site replicates the same structure and adopts the same terminology, communication methods, and management protocols so that different teams can work together seamlessly. This approach is reflected in the crisis management systems of many resource companies and is the basis of the approach described below.

CRISIS MANAGEMENT CYCLE

How a crisis should be managed is often presented as a cycle of mitigation, prevention, preparedness, response, and recovery. This Crisis Management Cycle, however, is an oversimplification that fails to communicate the most challenging aspect of crisis management. Crises are complex, dynamic, and non-linear. Response and recovery actions are seldom, if ever, applied in consecutive stages.

CRISIS MANAGEMENT SYSTEMS

This section outlines the essential elements of a crisis management system that could be considered representative of leading practices in the mining industry and that would likely meet the needs of most companies operating in the resources sector. These elements are:

- Crisis risk assessment based on operational, organisational, stakeholder, physical, financial, economic, political, and social factors;
- CMP including role-based checklists;
- One or several role-based CMTs;
- Crisis management software;
- At least one dedicated Crisis Management Room (CMR) and several contingency locations;
- Several special-purpose work areas close to the CMR;
- Comprehensive register of key stakeholders;
- Access to a range of communication systems;
- Competency-based training programme; and
- Annual schedule addressing documentation, training, and assessment needs.

Additional required services and support would include, as a minimum:

- Reliable broadband internet service;
- Site-based emergency response teams and equipment;
- Site-based medical staff and clinic facilities; and

- Access to crisis management consultancy services for facilitation of crisis simulations.

Additional requirements for many remote sites would include:

- Site-based security team;
- Medical staff and clinic facilities;
- Access to a country-based medical services provider;
- Accommodation facilities close to the CMR; and
- Emergency power supply for the CMR.

CRISIS MANAGEMENT PLAN

The CMP is an essential component of a Company crisis management system that documents the general approach, arrangements, resources, and procedures the Company has identified as requirements for managing crises.

When drafting a CMP, the starting point is recognising that the document's primary use is as a technical reference when developing training materials, updating role checklists, maintaining team capability, and other outcomes in support of crisis preparedness. When a crisis is underway, there may be little or no time to read a CMP. If the team has been adequately trained, there should be little need to do so.

In almost all situations, the following hierarchy of priorities should apply, starting with the highest priority: safety of employees and the public; protection of the environment; protection of Company reputation; protection of Company assets; and maintenance of operations.

A proven effective approach to managing Company-level crises is employing role-based CMTs. The team roles are aligned with the business's main functional areas. When members of the management team join the CMT, they are assigned one or several roles in accordance with their knowledge, experience, and personal qualities such as leadership skills. To allow for redundancy and shift handovers during an extended crisis, it is vital to have at least two team members available for each role, and for critical roles, at least three options are recommended.

A CMT's functions in managing a crisis typically cover five areas: emergency response, impact mitigation, business continuity, stakeholder engagement, and continuous improvement. A Company may delegate some of these functions to work groups acting in support of the CMT or specialist service providers, e.g., a media communications consultancy.

The roles of CMT Leader and CMT Coordinator deserve special attention. They are the only roles considered mandatory for all events, and the capability of the personnel assigned to these roles will be a key contributor to the performance of the team as a whole. Less than capable performers in these roles need to be reallocated to another role or dropped from the team as quickly as possible.

CMT Leaders are authorised to declare a crisis event and activate the CMT. During a crisis event, they are responsible for supervising the activities of the CMT and ensuring that critical outcomes are achieved.

The CMT Leader role demands strong to outstanding competencies in team leadership, problem-solving, operational planning, and interpersonal communications – a high tolerance for stress and robust self-belief is also useful.

During an event, the CMT Coordinators have only one primary accountability: ensuring that the other CMT members are effective in their roles and that management priorities, decision-making, and general procedures follow the requirements of the CMP.

The CMT Coordinator role demands systematic thinking, a strong background in management systems and operational controls for risk mitigation, sufficient assertiveness to challenge and correct team members irrespective of seniority, and a degree of indifference to interpersonal conflict or hurt feelings in the work environment. Less obviously, good coaching and training skills are essential, as the role often requires delivery of training and coaching of team members during actual events.

A well-proven workflow for a CMT is shown here (steps 12–17 refer to an ongoing work cycle that may be repeated many times):

Step	Action
1	Identify crisis event
2	Determine initial CMT composition
3	Establish Control Room location and time of first CMT meeting
4	Call out CMT members and ready the Control Room, including epidemic or pandemic precautions if necessary
5	Initiate an event in Crises Management Software (if available) and allocate team members to the event
6	Team briefing at the Control Room
7	Roundtable discussion
8	Notify other teams
9	Identify casualties
10	Identify key stakeholders
11	Identify critical issues
12	Develop response plans and assigning tasks
13	Assign breakout groups as required
14	Implement response plans and ongoing review
15	Commence "phones down" situation updates at an agreed frequency
16	Change out team members as required
17	Schedule CMT meetings
18	Declare crisis over and stand-down CMT
19	Secure records
20	Debrief team

A relatively recent development in crisis management practice has been the implementation of web-based crisis management systems. One example is Emqnet, which the provider describes as

a real-time issues and events monitoring, communication and recording application that can be accessed by selected teams or members of your organisation. It allows multiple teams, located globally, to maintain and share critical, real-time information on a central, viewable platform.

(www.emqnet.com)

Systems like Emqnet can dramatically improve certain aspects of managing a crisis event, including:

- Event initiation and assigning of CMT members to roles;
- Recording and managing information received by the CMT;
- Status reporting to Company management;
- Assignment and management of tasks;
- CMP document control; and
- Provision of a detailed and accurate record of management of the event.

On initiation of crisis response, the CMR becomes the working area for the CMT. The room should be equipped with the documentation, tools, and equipment required by the CMT in managing a crisis event.

Breakout rooms should be available within short walking distance of the CMR to provide work areas for breakout teams and individuals needing a quiet place to think, plan, or make calls to stakeholders. Separate media rooms and other rooms for affected employees and next-of-kin may also be required.

For crisis events likely to affect many stakeholders, a temporary call centre is recommended to avoid a high number of calls to the CMT and site management.

Stakeholder communications and media and social media management are integral to the crisis management process. The quality and timeliness of communications may be vital in determining the duration of a crisis and long-term losses to the Company managing the crisis, including reputation impairment. In a role-based CMT, the usual arrangement is that all stakeholder communications, including media releases, must be approved by the CMT Leader.

It is essential to have a detailed stakeholder register and know how to involve the right people before experiencing a crisis. Stakeholder mapping is then the visual process of laying out all the stakeholders potentially involved in a particular crisis on one map. The main benefit of a stakeholder map is that it provides a visual representation of all the people affected by the crisis or who can influence the outcome of the crisis and how they are connected.

Crisis communication is a specialised and extensively studied discipline, and many larger companies have staff specifically assigned to this area as well as access to the services of communications consultants.

Stakeholder communication is arguably the only area of crisis management that lies outside the typical competencies and experience base of resource industry management, and on occasions, it is spectacularly mishandled. Probably the most famous example in this regard is the Deepwater Horizon oil spill (2010). British Petroleum's (BP's) initial response to the crisis failed catastrophically. This was then compounded by basic mistakes in the communications to stakeholders (e.g., slow to properly acknowledge the families of the 11 workers killed in the rig explosion; or describing the environmental impact of the spill, the largest oil spill in the history of the USA, as likely to be "very, very modest"). The blaming of other involved parties by BP executives made the Company appear callous. Eight weeks into the crisis, BP's stock had nearly halved in value.

BP faced that crisis at the outset of the social media era, several years before "smart" phones became globally ubiquitous. The potential for rumours, misinformation, and disinformation to achieve "virality" in an hour or two has supercharged the news cycle and requires additional management capability in a crisis.

Right way to close an operation

All industrial enterprises eventually come to an end, at which time various closure procedures are required to rehabilitate the site and prepare it for subsequent use. It is important to distinguish between closure due to project completion which is usually well planned and closure due to financial distress which may be planned poorly, if at all. The measures needed for closure will depend mainly on the activities and processes at the site, the operation's scale and longevity, and the likely subsequent use of the land. The local environment and nearby communities' characteristics and legal and regulatory requirements will also be important in developing a Closure Plan. Mines and mineral processing facilities; oil and gas fields, storage facilities, and refineries; chemical plants; and power plants are among the industries where extensive and intensive closure measures are usually required. Agricultural and timber-based industries, although less problematic, may also require comprehensive closure measures. Waste treatment and disposal facilities generally are among the most problematic businesses or components of large sites.

Closure of an operation marks the laying-off or re-deployment of employees. Large companies with the advantage of multiple operations are usually well equipped to handle this process. Operations may be transferred to contractors' years ahead of closure, and these contractors transfer personnel from one project to the next, while the operating Company re-deploys its personnel within the organisation. Of course, there are often locally hired employees who have no interest in being re-deployed at another location.

Many business leaders have never had the experience of closing whole operations or dismissing large numbers of employees, a

DOI: 10.1201/9781003134008-15

distressing and challenging task. As a result, mistakes can be made. Typically, business leaders assume they must appear tough, deciding later to delegate implementation to others with the instruction to "Go fast!" This will seldom work. To be successful, business leaders need to balance "soft hands" and "hard hands" during closure and must stay involved. It is important to be decisive but also to be closely engaged in ensuring that employees, customers, suppliers, and communities are treated with consideration and compassion.

DELAY, CLOSURE, OR SALE?

In extractive industry sectors, producers sometimes delay abandoning oil and gas fields or closing mines, on the basis that future demand recovery or price increases will enable profitable operations to be resumed. The question then arises: is there a realistic prospect of resumption or is the delay intended to postpone or avoid decommissioning costs?

Before closure, there must be consideration whether the operation is to be mothballed (as occurs with many offshore oil and gas structures), permanently closed, or sold. When selling, different issues arise, such as timing and the terms for the buyer's protection from environmental liability issues. Sometimes, of course, site closure is required because all avenues for sale have been unsuccessful or further profitable operation is decidedly not feasible.

WORKFORCE REDUNDANCIES

In well-developed economic regions, a trained workforce is likely to find alternative employment after being laid off. In areas far from population and economic centres, closures of operations have more pronounced negative effects on host communities (BP 02). Indeed, the loss of a local economy's dominant economic engine can expose the fragility of a narrow economic base. As a result, such "mono-industry" towns and regions face multiplying and damaging impacts from closure. The loss of employment substantially reduces the flow of income through local economies, affecting retail, food services, other dependent sectors, and social services.

In remote operations, closure is in large part about mitigating impacts on people and communities. Closure measures should be underpinned by (1) dialogue with affected stakeholders to determine scope, scale, and timing of closure; (2) adequate planning from the outset; (3) provision of temporary income support to workers and their families, complementary to existing social protection programmes; and (4) deployment of active labour market policies that offer services, programmes, and incentives to encourage and enable re-employment among laid-off workers.

CLOSURE PROCESS

The environmental process for site closure broadly falls into five phases. Phase 1 involves collecting data, identifying stakeholders, and understanding legal and contractual obligations. Phase 2 requires extensive consultations with host communities. Their expectations will inform the requirements that the regulatory authorities will impose, including the required level of clean-up and restoration. Host communities frequently request that some of the existing infrastructure be retained for their future use. Such items include access roads, water supply, and power generating facilities. The expectations of those host communities that maintain traditional beliefs in respect of connection to the land can be more strongly influenced by spiritual considerations than rational or practical ones. Great sensitivity needs to be exercised, with strong involvement of Community Relations specialists when negotiating the closure requirements.

Phase 3 involves carrying out site investigations, characterising the site, and finalising any remediation and rehabilitation requirements, including negotiating and finalising the contracts. Phase 4 comprises the main closure tasks: demolition and removal of buildings, plant, equipment, and infrastructure; remediation of contaminated soil and groundwater; and site rehabilitation such as landscaping, regrading, slope stabilisation, and revegetation. The fifth and final phase is to verify closure work has been carried out to satisfy involved stakeholders and, ultimately, to finalise surrender and exit approvals (site relinquishment). This phase may require a period of decades.

CONTRACTUAL PROVISIONS AND GOVERNMENT REQUIREMENTS

Many industrial plants, mining tenements, and oil and gas Production Sharing Contracts are held on long-term agreements.

Until recently, such lease agreements seldom contained detailed provisions on how to relinquish a site that took into account environmental aspects. Commercial negotiations often define what is required in terms of site handover and clean-up, including budgetary funds allocated by the site user for potential environmental liabilities, often as closure bonds or guarantees. The landlord (in the extractive industry sectors usually the Government) will focus on ensuring it does not take possession of a site that contains pollution risk or other environmental, health, and safety liabilities and on maximising the value of the site in the future.

Sometimes, regulatory environmental clean-up requirements and expectations from the landlord differ, prompting lengthy negotiations to make the landlord's surrender requirements compatible with requirements of the environmental permit.

CLEAN-UP: KEY INFLUENCES

Remediation issues are discussed in BP 12. Two main questions asked by site operators considering closure are "What level of remediation will be needed?" and "What are closure criteria for site restoration?" Obtain legal advice regarding aspects of the property's environmental issues under the lease terms, to understand the strength of the negotiating position the landowner (or Government) holds. Another factor that impacts significantly on what needs to be done is how quickly we wish to close the site. A more balanced approach can often be taken if plenty of time is available in which to manage a clean-up and site rehabilitation.

However, if a Company wishes to or must exit a site quickly, it may be forced to carry out more complex and expensive remediation and restoration measures determined after negotiating from a position of relative disadvantage.

RELINQUISHMENT OF LEASE AND PERMITS

Many operations will have an operating agreement (e.g., Production Sharing Contract or Contract of Work) and environmental permits issued by the relevant Government authorities. A key issue when considering site closure is closeout and surrender of these operating and environmental permits.

It needs to be established whether permits need to be modified for closure and at what point they should be relinquished. An environmental licence often contains expressed conditions requiring notice of change in the process. If the operation is to proceed to closure, there is likely to be a formal relinquishment procedure. This process will dictate, to a large extent, what needs to be done with the site from an environmental perspective.

Before the Government accepts the lease agreement's relinquishment, it must be satisfied that the permit holder had taken all measures to avoid or eliminate any pollution risk resulting from the past operation. For the Government, returning the site to a satisfactory state often means reinstating the site to a condition similar to what prevailed before the permit was granted. This is not always wholly realistic and can lead to disagreement. On the one hand, the Company may hope for a swift and cost-effective relinquishment, while the Government will be reluctant to allow relinquishment until it is wholly satisfied that the stated environmental aims have been achieved. Governments like to err on the safe side. Some Companies have similar concerns, preferring to delay relinquishment until post-closure monitoring confirms that there are no ongoing environmental liabilities. While the operational and environmental permits are still in existence, regulators have an element of control through the sanctioning regime. The Government will wish to ensure the required clean-up and site restoration are achieved before it loses that control. Site relinquishment in the form of surrendering all permits takes considerable time in practice and is one of the critical issues to look at in planning site closure. In some jurisdictions, the process of relinquishment may be a well-trodden path, but in many it is not. Numerous developing countries only began to face issues of operational closure at the end of the last century.

Another area the Company should consider is whether it may be beneficial to maintain some operational permits. A water abstraction licence is a case in point, as the licence may have value in its own right, possibly in combination with residual infrastructure elements.

LONG-TERM CLOSURE LIABILITIES

Industries that have operated for many years may have very high closure costs. This is particularly the case where operations commenced before the implementation of environmental management measures. Closure in such instances can require expending hundreds of millions of dollars and may take decades to complete. In some cases of ongoing contamination, such as severe acid mine drainage, it has proved impractical to remove contamination sources. In such situations, the responsible companies have opted to manage the impacts by preventing the offsite spread of contaminants, involving interception and treatment of groundwater or surface drainage for many decades.

CLOSURE IN THE MINING SECTOR

Closure of mining and mineral processing operations can involve the most extensive, challenging, and expensive measures and procedures. Accordingly, much of the discussion in this Chapter is based on mining industry practice and experience. The mining industry has been active in addressing the issues and in developing guidelines for good practice. Detailed guidelines for mine closure have been prepared by the International Council on Mining and Metals (ICMM 2019). The Australian Government Department of Industry, Innovation and Science and the Department of Foreign Affairs and Trade jointly sponsored a handbook on Mine Closure as part of their Leading Practice Sustainable Development Program for the Mining Industry (DFAT 2016).

As well as the technical aspects of closure, these guidelines address the social issues involved, including unemployment and community disruption; opportunities for amelioration of adverse social impacts and creation of beneficial outcomes are also discussed. A suite of International Organization for Standardization (ISO) Standards for Mine Closure and Reclamation Management was prepared in 2020 by ISO Technical Committee 82 (TC 82/SC 7– Mine closure and reclamation management). The status of these at this writing was as follows (the first three are available for purchase at iso.org/committee/5052041/x/catalogue/):

- ISO 20305:2020: Mine closure and reclamation – Vocabulary: Published September 2020;
- ISO / DIS 21795-1: Mine closure and reclamation planning – Part 1: Requirements: Close of Voting;
- ISO / DIS 21795-2: Mine closure and reclamation planning – Part 2: Guidance: Close of Voting;

- ISO/AWI 24419: Mine closure and reclamation – Managing mining legacies – Requirements and recommendations: Under Development.

Most jurisdictions where mining takes place have promulgated mine closure regulations. Commonly, these include provisions whereby the mining Company must accrue funds to pay the costs of closure, such funds to be available even in the event of an unplanned closure. Closure costs can be extremely high. In the past, it was frequently the case that final costs of closure far exceeded the accrued funds. Provisions for funds accrual require closure costs to be estimated from the outset and revised periodically as the project proceeds. To provide the basis of the initial estimate, most jurisdictions require a Closure Plan be submitted and approved as part of the Environmental and Social Impact Assessment process, with regular updates throughout the project's operating life.

Experience indicates the importance of the Company and the regulatory authority agreeing in advance on which expenses can legitimately be allocated to the Closure Accrual Fund, as distinct from those deemed as regular operating expenses. Examples that have been challenged by regulators include the expense of consultants preparing and revising Closure Plans, laboratory testing charges, expenses involved in vegetation trials, and allocating portions of environmental and Community Relations staff salaries.

The first and most important closure activities are designed and implemented to ensure that the site is safe and will remain so. Shafts can be backfilled or capped while tunnel portals can be blocked. It is usually impractical to backfill open pit mines. Based on geotechnical studies, the most common approach is to stabilise slopes and prevent access to areas where hazards remain. High cost closure items are reshaping of waste dumps and Tailings Storage Facilities to facilitate surface drainage and control erosion. Control of Acid Rock Drainage, if required, can also be costly and may continue for many decades.

Many mines and operations in some other industries can reduce the time and effort required for closure by progressively rehabilitating areas where activities have ceased. This has the added advantage that spare operational capacity can be deployed when available, reducing the need for post-closure expenditure. Similarly, measures designed to ameliorate closure's social effects are also best implemented well before operations cease. Such efforts include capacity building and livelihood replacement programmes, such as agricultural extension, targeted training, and seed-funding programmes for alternative livelihoods and ventures.

CLOSURE IN THE OIL AND GAS SECTOR

The geology of the hydrocarbon reservoir, the nature and condition of the production facilities, and the production history all influence the environmental issues in decommissioning a specific property or installation. Despite the considerable differences from one field to another, common environmental concerns do exist, and examples follow: hydrocarbon and chemical releases; sumps and ponds containing oily residues; encrustations (scale) containing Naturally Occurring Radioactive Material (NORM); potential mercury contamination; presence of asbestos or polychlorinated biphenyls; potential for emitting hydrogen sulphide gas; disturbance to land and seabed and accumulated drill cuttings; and secure treatment and disposal of residual wastes. Age and size of the site will determine the extent, complexity, and cost of decommissioning.

Decommissioning (often used interchangeably with terms like "abandonment" and "recommissioning") is the process by which options for the physical removal and reuse or disposal of installations at the end of their working life are assessed and implemented at Cessation of Production (CoP). For new installations, the decommissioning process begins long before CoP; initial work is in fact part of the project design phase. For older installations, effective preplanning should commence at least 2 years before decommissioning. The two greatly divergent types of decommissioning in the oil and gas industry are onshore and offshore decommissioning.

Onshore decommissioning is largely uncontroversial and adequately covered by most national legislation. It involves well plugging and abandonment to protect groundwater, and removal of wellheads, flow lines, storage tanks, waste handling pits, and processing equipment. Most national legislation also covers site remediation requirements if soil and groundwater contamination is present.

By contrast, offshore installation decommissioning is a relatively new challenge to most oil and gas producing countries. The industry's experience in building platforms is naturally much greater than in dismantling them. Mature offshore oil and gas installations present unavoidable future issues as installed platforms reach the ends of their useful production lifetimes. Several different decommissioning options exist, and each results in an array of environmental and socioeconomic impacts, both positive and negative. These impacts are perceived and valued differently by stakeholders from different and contrasting perspectives.

Decommissioning differs widely from installation to installation, as the complexity of oil platforms varies widely. A "One Size Fits All" approach does not apply; neither is there a "Goldilocks" solution. At a minimum, decommissioning must comply with national legislation and international agreements and standards. Newer generations of agreements between States and petroleum companies treat the responsibility for decommissioning more effectively, while earlier generations typically did not. The earliest contracts often assumed production would continue under national management, and, if not, focused more on taking possession of reusable hardware than on residual site conditions.

Four broad strategies exist for decommissioning, with total removal being often the base case preferred by Government authorities. In determining which strategies to adopt for a particular installation, it is important to consider means to most cost-effectively minimise liability.

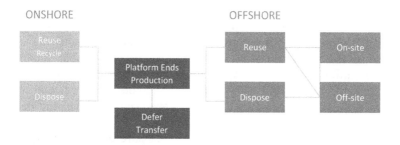

Leave in situ – The entire installation is left without any attempt to restore the site, an acceptable strategy if the installation is still in good condition, particularly if further production is feasible. Leaving in situ is an option that is only to be practised for deferral cases and is not an option for permanent abandonment. Deferral can be justified if, for example, continued operation of the installation under likely changes in market conditions is a possibility.

Transfer of Asset – This refers to transferring the offshore installation to other companies; liability for abandonment and restoration in such cases would need to be assessed and agreed with the new owners. Transfer of assets will only defer decommissioning.

Partial Abandonment – The degree of abandonment will depend on legislative requirements and may consider the environmental value of the area that has been impacted by operations. It is, however, common to strip out the topside hardware in partial removal.

Total Removal – Every component of the installation, both above and below the water surface, is removed, and the site is restored to its predevelopment condition to the extent practical. This option is likely to be most favoured in terms of practicability and acceptability.

OTHER INDUSTRIES

Most industries use hazardous materials such as fuels, lubricants, chemical reagents, pesticides, fire retardants, and cleaning compounds. Spillage or leakage of these materials constitute the most common causes of soil contamination and/or groundwater contamination (BP 12). Environmental Site Assessments (ESA) are generally focused on locations where these materials are stored as well as sumps, drains, and effluent disposal sites. It is necessary for old industrial sites to identify locations of potential historical contamination, requiring research into previous industrial activities. A Phase II ESA may be required where strong evidence of such contamination exists. This investigation may in turn trigger requirements for further study of the feasibility and design of remediation efforts or the assessment of residual risks needed for negotiations with regulators (further discussed in BP 12).

In designated industrial areas, the closure of one industry is usually followed by the establishment of another industry on the same site. In such cases supporting infrastructure may be retained and sold or otherwise transferred along with the property. Valuation and sale of the property are facilitated if the departing industry had already commissioned an ESA and carried out any necessary remediation. In such cases, prospective purchasers may appoint their own consultant to review the documentation and advise on any residual risks.

DOCUMENTATION

In the same way that "as constructed," as distinct from "as designed" documentation is required at the commencement of a project, detailed documentation of actual closure activities is important as it provides the basis for ongoing monitoring and, ultimately, relinquishment. This documentation should include details of the cleaning and disposition of plant and equipment; earthworks including details of what is buried and where – offsite or onsite; and the extent of any remediation including the processes adopted and the results achieved. An independent consultant's post-closure audit is also useful to demonstrate whether commitments in the Closure Plan have been met and the objectives achieved.

OUR ROLE AS BUSINESS LEADERS DURING CLOSURE

Be Visible and Personal – A closure is not an excuse for leaders to go into hiding. On the contrary, it is an occasion when employees, Government officials, communities, and other stakeholders need to see the commitment of senior business leaders. Leaders should set the tone and take personal responsibility for the organisation's behaviour during the closure process. This ensures that employees and business partners are being treated with dignity, fairness, and respect. It is instructive for business leaders to consider how they themselves would feel and how they would react if they were the parties affected by closure actions.

Address "What Does It Mean for Me?" – Employees and business partners have a right to be treated as adults and equals. Employees should be told why they are losing their jobs and how the closure will affect them (e.g., in terms of timing and severance benefits). Explain what you will do to help them and what you will need from them to help the Company, its customers, and Government regulators through the transition. Consider affected employees for opportunities at other Company locations or offer them contract work, particularly in extended closure efforts. At the very least, help them write effective résumés and learn how to leverage their networks.

Honour Company Commitments – During operations, all companies have made commitments to employees, customers, business partners, service providers, communities, and regulators. The Company should feel compelled to honour these commitments, or, if any of them change, stakeholders should be informed at the time- not when they are told their operation is winding down. Deliver messages that are consistent and positive but always grounded in reality.

Manage Closure Like a Project – If you develop an operation, you employ project-management techniques and appoint experienced project managers and contractors. The same level of planning should be applied to closures. Leaders must appoint an experienced, full-time project leader supported by a strong team, as the closure assignment will almost always be beyond the operations manager's role and experience. The project leader should have the authority and credibility to address the interests of all stakeholders: the Company, employees, customers, suppliers, communities, and regulators. The Company's reputation and image in mass media and social media will be damaged if closure activities are mishandled.

Looking ahead

In this closing chapter, it is appropriate to reiterate pertinent environmental trends that business leaders face in our society. Some trends, such as climate change, have now become entrenched and are very familiar; others may seem comparatively obscure, known primarily to specialists in that field. Nevertheless, the importance of environmental and social aspects of governance has morphed from vague discomfort among investors, customers, and consumers about environmental and social issues to the current situation in which there is genuine and serious concern or even alarm. Many stakeholders now view the global challenges of climate change, plastic pollution, biodiversity loss, water shortages, and economic and social inequality as tangible threats.

Social and environmental issues progressively affect stakeholder behaviour and the larger business environment. For a current example, nickel mining, a core industry in Indonesia for a long time, is growing rapidly to meet growing battery demand. A major environmental concern is the disposal of the resulting mine waste. Several companies proposed deep-sea tailings disposal. Regulators in Europe and elsewhere, worried that climate benefits from the global clean energy transition could be at the expense of marine pollution, threatened to ban stainless steel containing nickel from Indonesia. By the end of 2020, all Indonesian nickel producers abandoned their plans for marine tailings disposal, forcing them into onshore waste management programmes in high-rainfall, high-biodiversity, and seismically active regions. Another topical example is the difficulty, if not impossibility, of securing international finance for new coal-fired power plants or thermal coal mines.

DOI: 10.1201/9781003134008-16

MILLENNIALS

Millennials, labelled the most educated and knowledgeable generation in history, prefer healthy food to fast food. Millennials, strong on saving for their future, prefer to invest in organisations with a positive influence on the environment. They put their money where their values are. They use technology in various aspects of their lives to support diversity and inclusion and believe they will make the world a better place.

MILLENNIAL   GEN X   BOOMER   SILENT
Born 1981–96   Born 1965–80   Born 1946–64   Born 1928–45

Millennials, people born between 1981 and 1996, comprise the most numerous subset of the world population and are now the workforce's largest demographic. They have different traits than other generational cohorts, including most of today's business leaders, the ageing Gen X and Boomers. Business leaders are well-advised to embrace the changing trends in outlook, consumer behaviour, and the business environment, as millennials' spending power will only increase while older generations leave the workforce and reduce consumption. Millennials will inevitably replace their elders as they move into senior positions, changing Company cultures, especially with regard to Environmental Social Governance (ESG).

Generation Z, people born from 1997, are consumers and have been entering the workforce since about 2015. Any implicit assumptions that they will mirror millennials in attitudes and behaviour remain to be demonstrated.

GENDER AND DIVERSITY

Pch.vector

Despite a broad focus on diversity and inclusion over the past two decades, women are still vastly under-represented at all levels in most industries. It is true that, going into 2020, women held the majority of jobs in the USA for the first time in almost a decade. But this has not translated into meaningful progress at the corporate level, where the proportion of women has hardly changed. It may take another 135 years to close the overall global gender gap, according to the World Economic Forum's 2021 Gender Gap Report. This cannot be an excuse for today's business leaders to ignore gender diversity in their companies. So what are the obstacles?

To give the benefit of the doubt, few men set out to discriminate systematically against women; their unconscious bias quite effectively does it for them. Hence recognising and addressing unconscious bias is the first step toward building a business culture that achieves gender diversity and can deliver benefits by enabling all employees to develop their full potential. Beyond that essential step, businesses need to create an environment in which women can count on support in the same tangible ways that men have always taken for granted. Women need executive sponsorship and mentorship, just as men do, and succession planning needs to bring women into the picture. This seldom happens, even in companies with a strong pipeline of high-performing female candidates.

Yet gender inclusion often proves easier to achieve than bringing other excluded and marginalised groups into the fold. Racial, religious, and social minorities need all the same considerations essentially as outlined above. If good intentions and simple justice cannot make the business case for this, consider the overall purpose of this book – helping leaders understand how to achieve resilience to survive in the face of massive, rapid, and unrelenting change. Just as a century and a half of natural science showed clearly that diverse ecosystems successfully adapt to disruption, modern management science has demonstrated that more diverse organisations better anticipate, cope with, and adapt to change. This is simple fact, not a matter of "political correctness." To achieve equality of opportunity, gender and otherwise, good intentions should lead to concrete actions. Take pride in a leadership role at a Company that respects, values, and embraces the benefits of peoples' differences. Work toward the day when a lack of diversity is a relic of the past and the "old boys network" refers to a group of men who have long since retired.

ENVIRONMENTAL AND SOCIAL ISSUES ARE THE NEXT PHASE OF GOOD GOVERNANCE

Forget for a moment how management can affect sustainability. How will sustainability change management? Business leaders can change the future of their companies by merely changing their attitudes toward ESG. Surely, traditional corporate governance

will remain important – particularly around initiatives to improve Board quality, employee benefits, shareholder rights, and management incentive structures – but the governance of environmental and social aspects will more and more become a focus for investors and Boards. Expect boardrooms to be ever more concerned with ESG issues as companies attempt to respond to both the economy and the vicissitudes of an industry in a fast-changing world.

Companies that ignore this trend – whether by disregarding diversity and inclusion practices, overlooking environmental and social issues, or acceding to unethical employee behaviours – will have a higher risk profile, and this will be quite evident to all interested stakeholders. These companies will be less attractive for employees, customers, and investors. Some institutional investors have gone so far as creating ESG teams with authority to veto investments in companies if perceived ESG risk is significant or if the potential target will not commit to widely accepted ESG principles.

ESG has its own algorithm: responsibility, accountability, serendipity, and epiphany in some unknowable combination. The time and effort required to systematically scrutinise a Company's CSR efforts to integrate sustainability as a tool to manage risk and create long-term shareholder value will be most beneficial. Understanding and evaluating the Company's impact on the environment will be a valuable skill at the Board level. It will become a sought-after addition to many Boards.

ACCELERATED CLIMATE ACTION

Climate change will remain a dominant theme for years to come, as Governments worldwide introduce more climate-related regulations,

and will be reinforced by the policies of investors, financiers, and insurers. Expect time-bound commitments to net-zero GHG emissions by companies and investors to become standard practice. All sectors – including those that are emissions-intensive and have shown resistance in the past – will transition to the low-carbon economy as companies recognise the risks linked to ignoring climate change as well as the opportunities and competitive advantages in proactively addressing these risks. The perception that anthropogenic emissions will lead to damaging climate change has become entrenched and can no longer be ignored by business.

Many companies will aim to seize new business opportunities and position themselves as climate leaders. At the same time, investors will increasingly incorporate climate risk into their investment and shareholding policies, to include voting against the Boards of laggard companies. "Do not let what you cannot do interfere with what you can do" (*John Wooden*). Is the Company affected by climate change? What steps does the business take to address those effects? Does climate change create business opportunities?

Even for those who believe that the risks of catastrophic outcomes from climate change are overstated, it would be most unwise to ignore the pressures to reduce GHG emissions. There is nothing to be achieved by swimming against the current. If, in the future, it becomes clear that the risks are less than previously perceived, then the current will change direction, and responses can be changed accordingly.

INTERNET OF THINGS AND BIG DATA

Data and technology drive significant changes in measuring, calculating, and monitoring ESG factors and assessing their materiality and impact on long-term value creation. New technology and data-management systems, e.g., the Internet of Things (IoT), have only increased the collection and analysis of massive data sets or "big data."

Big data enable a business to discover patterns and relationships that could not be unearthed with traditional approaches. The IoT market may be in its early days but is already a vital enabler for economic growth and sustainable development.

Big data will reshape environmental management. Networks of sensors to collect environmental data, such as air quality or water consumption, and big data visualisation methods, will help business

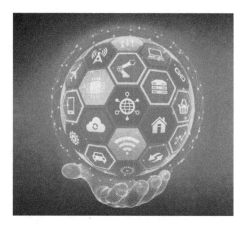

leaders link causes and effects. Big data analytics also help businesses optimise resource use. For example, networks of sensors can collect water and energy use data, allowing companies to see where they may be inefficiently spending resources or incurring risks. Advanced systems can automatically adjust buildings to optimise energy use, e.g., by deactivating lighting when a room is not in use.

At a different scale, Earth observation big data derived from satellite sensors increasingly help understand and predict natural processes, human impacts, and the current and future state of the environment. Regularly collected remote sensing and in situ measurements will identify trends and provide accurate and up-to-date knowledge about environmental change, invaluable information for any land-intensive industry.

ENERGY MANAGEMENT

Industries consume considerable amounts of energy. Surely all businesses must continue to commit to an overhaul in favour of greater sustainability by reducing energy consumption and, along the way, energy costs and CO_2 emissions. The low energy consumers will have a competitive advantage in the face of energy price volatility and carbon pricing.

Apart from investing in more environmentally friendly machinery, switching to more energy-efficient operating systems is a popular energy management option, both for companies looking

to update old systems and those trying to reduce their carbon footprints. Managing energy consumption is an ongoing process, first to spot easy-to-fix faults, and second to find new energy-saving opportunities that can be implemented and monitored to track progress achieving ongoing energy savings.

As businesses are forced to become more environmentally conscious, companies look for ways to curb energy use. An intermediate step has already commenced, whereby many hydrocarbon fuels are being phased out in favour of more efficient hydrocarbons (i.e., less CO_2 emissions per kW generated). Increasingly, hydrocarbon combustion processes will be replaced with ammonia, hydrogen, or electricity, but these only represent progress if generated by alternative renewable sources like wind, water, or solar. The share of renewables in electricity generation had jumped to nearly 30% by 2020, mainly through replacement of coal and gas. However, those two sources still represent more than 60% of the global electricity supply.

The years ahead promise further growth in the renewable energy sector, whether due to increased innovation or increased collaboration among business stakeholders, including regulators. Renewables also grow as utilities and regulators prefer them to replace retiring capacity; customers switch to renewables to save costs and address climate change concerns. Growth in renewable energy creates opportunities for most companies and threats to conventional energy producers. However, companies ready to innovate, collaborate, and seize new opportunities, including the traditional energy producers, will likely thrive in a new renewables growth phase.

Renewables projects require capital investments in physical infrastructure that inevitably have their own environmental impacts, including displacing competing land uses, habitat destruction, wildlife mortality, and, ultimately, closure with its attendant waste generation. It follows that there will be continued incentives for further innovation and new technologies if the world's future energy needs are to be met. Downside risks, including environmental risks, are likely, whatever energy sources are developed in the future.

RELATIONSHIP TO PLASTIC

Since the early 1950s, plastic has found its way into almost every aspect of human life, mainly due to its versatility and low production cost. To put it in perspective: about 20,000 single-use plastic bottles are purchased every second, more than half a trillion per year, creating an environmental threat that some activists predict will be as severe as climate change. The problem is that plastic may take hundreds of years to decompose – if ever it does.

Most people are already familiar with the dangers of plastic, its resistance to biodegradation, and its effects as a marine

pollutant. Companies are beginning to respond to plastic's negative environmental impacts and to seek ways to reduce its use. For example, Coca Cola, producing annually more than 100 billion plastic bottles and among the biggest creators of plastic waste, has pledged to recycle as many plastic bottles as it uses by 2030. This trend toward plastic recycling and biodegradable products can only increase with Governments everywhere beginning to enforce complete bans on single-use plastics.

> If you cannot do great things, do small things in a great way.
> *(Napoleon Hill)*

Most companies can implement simple measures to great effect. In mining operations, especially remote sites, providing drinking water can be logistically challenging and costly, accounting for as much as 10% of the catering costs. Single-use plastic water bottles are still the most common choice. Yet the switch to multiple use water containers is relatively easy, reducing plastic waste significantly. Business leaders can decide if their Company is seen as friend or foe in the war on plastic. This challenge is not trivial and will not readily disappear.

ACCESS TO WATER

Water conflict is not new: Historical records suggest humans have been fighting over water resources for at least 5,000 years. And it

may become worse – growing populations and booming industries have made the world increasingly "thirsty" while a warming climate increases evaporation. Will the quest for water in this century be as economically important as the contest for oil in the last century, a factor in many wars? Operating a business requires an environmental licence and an enabling environment. Accordingly, access to clean and plentiful water supplies is a long-term business challenge. Multiple challenges will emerge in the years ahead due to climate change, rising customer demands, urbanisation, and the implementation of emerging digital technologies. Now is the time to question today's water practices, to reflect on which obstacles are approaching, and how business leaders can turn these challenges into opportunities that benefit the business, customers, and, perhaps most importantly, the environment.

NEW TECHNOLOGIES

New technologies, and the development of new applications for existing technologies, promise significant improvements in many aspects of ESG. Industry can assist in these developments through sponsorship and early uptake of applicable technologies. Major research is under way to develop a steel production process that does not require the use of coal. Similarly, a new process is being developed for producing aluminium with much lower power consumption. Despite the fact that cyanide is easily and safely managed, there is widespread opposition to its use for gold extraction, and its use for this purpose has been banned in some jurisdictions (e.g., Puerto Rico and the U.S. State of Montana). Accordingly, it is likely that an economic process for extracting gold without the use of cyanide will finally become available. Promising new energy sources are also being developed including small-scale nuclear reactors, enhanced and advanced geothermal systems, and hydrogen fuels and fuel cells.

Drones appeared in the 1940s, but their widespread use began not very long ago. Drones are now used for a number of applications such as delivery of emergency supplies, visual monitoring, and surveying. Future applications are likely to include various types of environmental sampling, especially in areas that are difficult to access.

Also developing rapidly are numerous types of sensors that provide industry with accurate, real time information about processes, energy use, emissions, air quality, water quality, and machinery diagnostics.

BIODIVERSITY AS AN EMERGING MATERIAL RISK

Biodiversity can affect a company's ability to secure project finance and the licence to operate; it can disrupt the supply chain and damage reputations. Tightening societal expectations are being followed by increasingly stringent legal compliance requirements, including baseline studies, impact assessments, and mitigation and offset measures. However, biodiversity is an economic externality

difficult to measure, and its relationship with the corporate sector remains poorly understood.

Investor attention to biodiversity has to date primarily focused on the extractive industries, assuming that biodiversity is most likely to be material in these sectors and because biodiversity impacts of land clearing associated with mining are more easily understood. However, biodiversity risks are related to many other sectors, including food production, forest industries, and water supplies. Effects of climate change and the plastic waste on biodiversity are only starting to be understood.

Looking forward, all companies need to demonstrate high standards for biodiversity to reduce their risk exposure. Indeed, safeguarding shareholder and natural values are not mutually exclusive – they are interdependent. A Company potentially exposed to high-biodiversity risks should assess the ecological context of the issues. If these risks are material to the business, they must be faced as a matter of critical importance. A biodiversity risk assessment needs to be followed by measures to manage material risks, threats arising directly from the Company's activities, and those arising indirectly from supply chains or the mismanagement of biodiversity by others. Uncontrolled forest clearing following access provided by geothermal or mining projects is an example of the latter risk and can be as much a social as an environmental issue.

Management of biodiversity is best integrated with the business's risk management systems and performance targets, complemented by public reporting on biodiversity management as documented in Annual Sustainability Reports.

EXTENDING GREEN PRACTICES ALONG THE SUPPLY CHAIN

Supply chain management has traditionally been viewed as a process whereby raw materials are converted into final products, and then delivered to the end-consumer. Since the early 1990s, companies have been faced with pressure to address environmental management throughout their supply chains. Rising environmental concerns increasingly push companies into greater awareness of their purchasing decisions. However, adding "green" concepts to supply chains seldom proves straightforward.

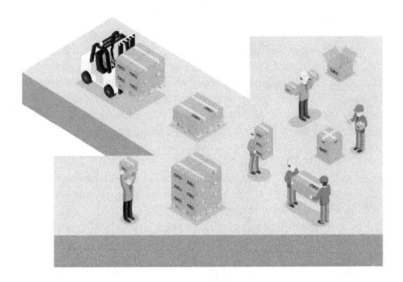

A sustainable supply chain management strategy will require companies to adopt environmentally friendly purchasing practices, including selecting materials composed of less ecologically harmful elements, fewer raw materials, and more renewable and recyclable resources to deliver to the end-user. Besides the stance of business leaders, external and internal pressures will decide the extent to which businesses change their supply chain management to improve sustainability.

This not only involves developing a greener supply chain. Other related considerations enter the equation, such as selecting low-carbon, socially, and ethically responsible suppliers. Social supply chain is the term used for supply chains that accept a trade-off between economic goals and social responsibilities to improve shared values with business stakeholders. An ethical supply chain focuses on the need for corporate social responsibility, working to produce products and services in a way that treats its workers and the environment ethically.

The Modern Slavery Act 2015, an Act of Parliament of the United Kingdom designed to combat labour servitude, is an example of legislative pressures to improve the ethics of supply chain management. The "Transparency in Supply Chain Provisions" require larger businesses to publish annual statements to confirm the steps taken to ensure that slavery and human trafficking do not exist in their business or supply chain. Alternatively, companies must declare that no steps have been taken to check the existence of slavery or trafficking, an option few businesses opt for, as it clearly places their ethics into question and will affect their reputations.

The 2021 Xinjiang Sanctions, following the USA blacklisting of 87% of China's cotton crop – one-fifth of the world's supply – citing human rights violations against Muslim Uighurs in China's north-west Xinjiang region provides another example. What was happening then in the fashion industry is rare in the history of global trade: a multibillion-dollar supply chain disintegrating almost overnight over a human rights issue.

SUSTAINABILITY AS COMPETITIVE ADVANTAGE

The business implications of sustainability merit scrutiny – and scrutiny of a different kind than that provided by the radical environmental activists. Can adopting significant sustainability practices lead to superior financial performance, or is it merely a strategic necessity to ensure corporate survival? Or worse, will sustainability practices damage profitability?

There is evidence that companies and investors who fail to act on ESG issues will gradually face greater business risks. They also miss out on significant opportunities available to ESG leaders, ranging from better access to capital to operational improvements. Demonstrating leadership in ESG is a differentiating factor for businesses, with business leaders having much to gain from embracing ESG stewardship as a competitive advantage, including superior financial performance. A firm move toward ESG is a valuable differentiator between your Company and the competition.

Most companies have already laid the foundations for addressing many of the trends and issues discussed in this chapter and throughout this book. Nevertheless, business leaders must expect to witness demand for more widespread and rigorous implementation of ESG-related practices across industries and jurisdictions in the years to come. They can ill-afford to continue being trapped in a Welchesque world and expect profit to remain the only metric of delivering true value. Expectations will increase. How will you respond?

Appendix: managing catastrophic risk – tailings disposal

Unwanted events, characterised by very low likelihood but extreme consequences, comprise a special class of risk known as Catastrophic Risk. A range of technical and human factors makes management of these risks uniquely challenging, especially at the enterprise or Company level, and the consequences of failure can be devastating for both the risk owner and the broader community. Failure to manage such risks across an entire industry may impact society at large and impede progress toward the goal of sustainable development.

For a range of technical and human factors, Catastrophic Risks are among the most challenging of risks managed at the corporate level. They lie at the limits of enterprise risk matrices and outside of the experience base of the teams assigned to manage enterprise risk. The rarity of these events across an industry makes it difficult to estimate the likelihood of a similar event for an operational site. The typically complex chain of causation associated with such events presents challenges in implementing controls for risk mitigation.

The disposal of mine tailings in a tailings storage facility (TSF) is associated with greater environmental and societal risk than any other mining aspect. It can only be described as a Catastrophic Risk. This risk has been the subject of numerous case studies, independent reviews, industry databases, and technical analysis over decades. It arguably is the most studied Catastrophic Risk being managed within the resource industries.

The purpose of this Appendix is to provide an overview of approaches for the management of Catastrophic Risk at the enterprise level using tailings disposal as a well-studied example. Although the context is tailings disposal, the principles of risk mitigation described in the pages following are equally applicable to the management of any complex operational risk within the resources industries.

CATASTROPHIC TSF FAILURE

As a lead-in to the following sections, it is useful to define the characteristics of a catastrophic TSF failure. For this purpose, it is fitting to reference the International Council on Mining and Metals (ICMM) Global Industry Standard on Tailings Management (GISTM 2020). This is the first and only international standard addressing the safety of TSFs, released at the time this book was being drafted. This standard defines a catastrophic TSF failure as:

> ...a tailings facility failure that results in material disruption to social, environmental and local economic systems. Such failures are a function of the interaction between hazard exposure, vulnerability, and the capacity of people and systems to respond. Catastrophic events typically involve numerous adverse impacts, at different scales and over different timeframes, including loss of life, damage to physical infrastructure or natural assets, and disruption to lives, livelihoods, and social order...

Tailings disposal practices

Tailings are a by-product of mineral processing at most mining operations. At these sites, crushing and grinding of the ore is followed by the addition of water and process reagents to produce a slurry. Following the recovery of the minerals of value, the slurry exits the process plant as tailings and is usually disposed of in TSFs or similar containment structures (Spitz and Trudinger 2019). It is estimated that there are more than 3,500 TSFs worldwide, some of which are among the largest of human-made structures.

Their design, construction, and operation vary; however, the factors that determine long-term performance and risk of failure are common to most facilities.

By design, a TSF will eventually contain large quantities of potentially hazardous or toxic material (stored volumes of 50 million cubic meters or greater are not uncommon). Under certain circumstances, such as earthquake shaking, stored tailings may be prone to liquefaction, marked by a sudden loss of strength and transition to liquid-like behaviour. Under conditions such as these, the impacts resulting from containment failure can be catastrophic; massive volumes of the contained tailings can flow through a breach within minutes.

Construction methods

The majority of tailings storage facilities feature a containment embankment constructed from the waste rock (mostly), from "Borrow Pits" specifically developed to produce embankment materials, or from the tailing material itself. This embankment may run between ridges to close off a valley or, if the topography is flat, forms all sides of the containment structure.

The embankments of TSFs are typically raised in stages over the life of the mine to provide storage for ongoing tailings production. Construction is staged because it utilises waste rock or tailings produced by the mine, and spreading construction over the life of the mine reduces up-front cost. This approach brings with it some consequences seldom seen in the construction of other large structures; the team responsible for the original design is often no longer involved at completion, and those that are still working in the industry may not have access to data on the performance of the completed structure. It also complicates the issue of accountability for design performance.

The embankments of TSFs are raised by one or a combination of three main methods known as upstream lifting, centreline lifting, and downstream lifting. The method used has important implications for the stability of the structure.

•   Upstream lifts are most often constructed using hydraulically deposited tailings, with each successive lift supported by previously placed tailings.
•   Downstream or centreline lifts are typically constructed largely from the waste rock, with each lift being supported by the embankment rather than by the tailings beach. Downstream and centreline lifting facilitates the construction of the embankment in different zones comprising rock, sand, and clay that specifically address mass stability, minimisation of seepage, and control of water levels within the embankment.

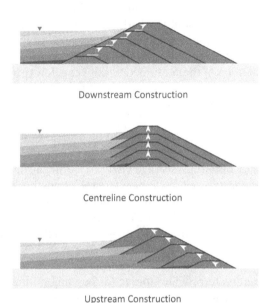

Downstream Construction

Centreline Construction

Upstream Construction

Risk of failure is generally highest for upstream construction and lowest for downstream construction.

Unsurprisingly, dams built by the downstream or centreline method are generally much safer than those built by the upstream method, particularly when subject to earthquake shaking (Chambers and Higman 2011).

The recent catastrophic TSF failures at iron mines in Brazil (Samarco in 2015 and Feijão in 2019) involved embankments constructed from tailings using upstream lifting. In both cases, this method of construction was directly linked to the failure. As of 2020, three countries had banned upstream lifting (Chile, Peru, and Brazil). Any Company committed to minimising tailings storage risk would have cause to reject designs based on upstream lifting.

Placement methods

Most tailings are discharged into a TSF as a slurry. However, a range of placement techniques is practised. Some of the water may be removed before placement to produce a paste, while removal of most of the water will produce so-called "dry tailings," which can be trucked or conveyed to a TSF and compacted after placement to increase strength. Dry tailings placement is a recent development offering significant opportunities for reducing the risk associated with tailings storage, especially in terms of liquefaction in the event of embankment failure, but may be challenging to implement in high rainfall locations, and generally imposes additional costs on the operation.

Performance objectives

The performance objectives for a TSF are usually established at the beginning of the design process. Common examples are:

*   Secure containment of all tailings and process solution during operations and after closure;
*   Control of acid mine drainage;
*   Final structure that supports site closure objectives;
*   Regulatory compliance; and
*   Cost.

Control of acid mine drainage is sometimes overlooked, even though uncontrolled oxidation may significantly alter the physical properties of waste rock used in embankment construction. The greatest risk in relation to all these outcomes is catastrophic embankment failure.

TSF failures globally

Between 1970 and 2020, at least 63 major TSF failures were recorded, and the frequency of high-consequence failures is rising (Owen et al. 2020). The impacts resulting from catastrophic failures can be dire. Three notable examples in the last decade include:

*   Brumadinho tailings dam disaster in Brazil in 2019, resulting in at least 134 fatalities (252 persons remain unaccounted for) and the release of 12 million m^3 of tailings.
*   Bento Rodrigues tailings dam disaster in Brazil in 2015, resulting in the release of 60 million m^3 of tailings and considered one of the worst TSF failures in terms of environmental impact.

- Mount Polley tailings dam disaster in Canada in 2014, resulting in the release of 10 million m³ of water and 4.5 million m³ of tailings.

In response to the Brumadinho disaster, a project to develop an international standard for TSF management was commenced in 2019 with support from the ICMM, the United Nations Environment Programme (UNEP), and the Principles for Responsible Investment organisation (PRI). The Global Industry Standard on Tailings Management referenced at the beginning of this Appendix began to be released in 2020.

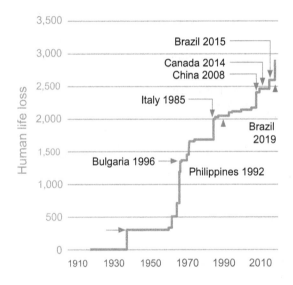

Source: Santamarina, Torres-Cruz, & Bachus, 2019.

WHY TSFs FAIL

Models of accident causation in widespread use across the resources industry may be usefully applied to the analysis of TSF incidents. The following key principles applying to operational incidents in general also hold true for TSF incidents.

There is rarely a single cause of an incident. A number of causes are invariably involved that in combination result in the unwanted event; a chain of causation is always involved. Proximate or immediate causes that directly contribute to events at the time of the incident always result from a chain of precedents leading back to root causes. It is the basic causes of an incident that must be addressed by corrective actions to minimise the risk of reoccurrence. Almost always, the basic causes are organisational factors.

What are initiating events, proximate causes, and basic causes, terms which are often confused?

Initiating events

Failure of a TSF may be associated with an initiating event or trigger such as an earthquake or an extreme rainfall event. In these cases, it can be argued that unsafe conditions were present leading up to the event and were the true cause of the failure. Earthquakes and extreme rainfall events are unavoidable, to a degree predictable, and must be allowed for in the design process.

Proximate causes of failure

The proximate causes of TSF failures are always geotechnical in nature. The most important of these are:

- Excessive amounts of water held within impoundment or embankment itself;
- Inadequate freeboard;
- Absence or failure of spillway;
- Inadequate embankment strength; and
- Inadequate foundation strength.

Excessive amount of water held within the impoundment is the most common proximate cause of failure, increasing the risk of overtopping, internal erosion, and embankment failure and tailings liquefaction. Earthquakes are of less consequence for most non-upstream tailings facilities.

Basic causes of failure

Review of several international databases of TSF failures has shown that all failures on record have involved elementary engineering or operating shortcomings. In every case, there was a lack of design ability, poor stewardship (construction, operating, or closure), or a combination of both (Davies 2002). This indicates that the basic causes of failures are almost always organisational factors associated with inadequate risk management. At some point leading up to a failure, decisions were made that compromised the safety of the facility.

This finding supports a corollary that TSF failures are preventable. The knowledge and skills necessary for this are readily available. What remains to be done is to ensure the mining industry consistently applies leading practices in engineering, operation, and closure of these structures.

Lessons learned from databases and case studies do not extend to the final stage of the tailings dam lifecycle, which is closure. Experience of the long-term behaviour of tailings dams after closure is limited. However, it is realistic to expect that all embankments, in the absence of ongoing monitoring and maintenance, will eventually fail due to erosion and other processes that will progressively reduce the factor of safety.

There is no standard applying to the required life of closure designs. However, a target of at least 1,000 years is gaining acceptance internationally. This may be an unrealistic expectation in many cases, particularly in higher rainfall areas. A more realistic goal could be closure structures that fail progressively over an extended period such that the rate of material released from the structure falls within the carrying capacity of the receiving environment.

ESTIMATING CONSEQUENCES OF FAILURE

Estimation of the likely consequences of failure is an important step in the design process for any TSF. This is approached in two main ways, described as follows.

Dam break studies

Loss of containment of tailings may involve several release mechanisms. An initial flood wave of water may be followed by a more gradual release of tailings solids through slumping, or all contents may rapidly flow out in the event of tailings liquefaction. Dam break studies model these processes to estimate the total area to be directly impacted by released material, called the zone of influence.

The zone of influence is an important input into the determination of the Consequence Category for a TSF (see below). It is also an important input for the development of a TSF failure emergency response plan, as it will indicate the likely scale of impact, the number of people and dwellings potentially affected, and required emergency response resources. Every TSF design report should include a dam break study.

The methods and assumptions applied in dam break studies for TSFs vary greatly, and the results are seldom validated against actual events. While establishing a zone of influence is an essential step in the design process, it should not be treated as necessarily accurate in the sense of an operational map.

Consequence Category

The Consequence Category of failure of a TSF is determined based on likely impacts in the event of a catastrophic failure (refer to ICOLD guidelines). Criteria include population at risk, infrastructure loss, and environmental impact. Due to downstream populations at risk, most TSFs would fall into the "Extreme" Consequence Category.

Consequence Category provides a systematic and transparent basis on which to:

- Rank the suitability of potential tailings facilities sites;
- Select appropriate design performance criteria;
- Determine operational requirements such as frequency of inspections; and
- Prioritise resources for the management of TSFs across multiple sites.

Consequence Category does not consider the likelihood of failure, and so in itself is not a measure of risk. The likelihood of failure is the subject of the next section.

ESTIMATING LIKELIHOOD OF FAILURE

Estimating the consequences of failure for a TSF is relatively straightforward, at least at a high level. Estimating the likelihood of such events is much more problematic. Every TSF is unique in terms of its design, construction, operational management, and site factors, and so the direct applicability of failure rates experienced across the industry to a particular facility is very limited.

Beyond direct comparisons, more sophisticated methods are available for estimating the likelihood of failure. *ORE2-Tailings software* (Riskope Consulting Firm) models the likelihood of failure for a given TSF based on 30 diagnostic parameters and allows benchmarking of the facility against a worldwide portfolio of tailings structures and recent failures.

PRINCIPLES OF RISK MITIGATION

Before discussing the management of risk associated with tailings facilities, it is useful to review several risk management principles that are generally applicable to the management of all operational risks, namely the precautionary principle, As Low as Reasonably Practicable (ALARP), and the hierarchy of risk controls.

< Precautionary Principle – When managing a catastrophic risk, it is often the case that Current Risk and the benefit available from additional, often costly risk controls cannot be quantified with any precision. In these circumstances, the precautionary principle can provide a basis for decision-making in the absence of a cost-benefit analysis or similar financial justification that is normally the basis of an approval to spend Company funds. One author has had the experience of working for a mining Company where the CEO directed that the goal of zero risk had replaced the corporate goal of minimising the risk associated with a site TSF. All opportunities for progressive risk reduction had to be considered for implementation (and this was for a site already implementing what was regarded at the time to be industry-leading practices for managing TSF safety). This was an excellent example of the precautionary principle in action.
< ALARP – Application of ALARP in the context of TSF risk was first referenced in ANCOLD guidelines issued in 1994. When a risk is described as ALARP, the cost involved in reducing it further is judged to be grossly disproportionate to the benefit gained. ALARP recognises that operational risks can never be reduced to zero and that it would not be in the interests

of society to require companies to incur huge costs in attempting to do so. ALARP is a useful concept to apply to TSF risk, especially when engaging with stakeholders. It does have a significant limitation in that the risk of failure associated with a properly designed and operated TSF is likely already low and difficult to quantify, as is the incremental benefit of additional risk controls. In such circumstances, ALARP becomes a matter of judgment. There is no doubt that risk associated with a TSF treated as acceptable by Company executives may be seen quite differently by local communities located within the facility's zone of influence.

< Hierarchy of Risk Controls – The hierarchy of risk controls is a widely recognised framework for risk and hazard control, especially within occupational safety management circles. It provides a basis for identifying and prioritising risk controls by ranking approaches to risk mitigation from highest to lowest in terms of effectiveness. The higher a control is in the hierarchy, the less it relies on human intervention or behaviour for success. The important risk controls applying to the operation of tailings facilities are engineering and administrative controls. Still, it is essential to recognise that there are options available from higher levels within the risk control hierarchy, including the substitution of TSFs, and the elimination of the need for tailings storage. These options are discussed in the following section Alternatives to Use of Tailings Storage Facilities.

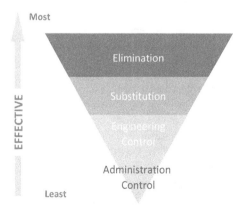

EXAMPLE OF RISK CONTROL

- Banning or restricting mining
- Deep-Sea Tailings Placement
- In-pit or underground tailings disposal
- Reprocessing of tailings with placement in the pit

• Site selection criteria	• Dry tailings placement
• Leading practices in design and construction	• Co-disposal
	• Storm freeboard allowance
• Design criteria based on Consequence Category	• Emergency spillway
	• Water treatment plant
• Risk assessments	• Control of water regime
• Annual risk mitigation plans	• Condition monitoring
• Independent expert reviews	• Preventative maintenance
• Compliance inspections	• Management KPIs
• by regulator	• Records management
• Regular status reporting to BoD	• Change management
• Operations Manual	• Staff competency
• Sub-aerial deposition	• Emergency preparedness

ALTERNATIVES TO TAILINGS STORAGE FACILITIES IN-PIT PLACEMENT

At some sites, it may be practicable to dispose of tailings in mined-out pits rather than in tailings dams, eliminating the risk of failure of containment. In practice, this requires the site to have more than one pit, and in any case a TSF is required to store tailings produced by processing of ore from the initial pit. The sequencing of mining between multiple pits is normally designed to meet processing requirements and maximise economic value. Although attractive in concept, in-pit placement of tailings is seldom implemented.

Deep-sea tailings placement

It is estimated that over 99% of mine sites that produce tailings utilise containment structures for their disposal. However, in some circumstances, the need for land-based containment can be avoided through techniques known as submarine disposal or deep-sea tailings placement (DSTP).

By discharging tailings to the ocean, the risk associated with tailings dams and similar structures is eliminated. If the sub-sea deposition area is sufficiently deep, the tailings will reside permanently in a zone of very low biological activity. Depending on the chemical characteristics of the tailings, placement in less deep locations may be viable. In these cases, the environment risk will always be insignificant compared with the risk over perpetuity associated with a decommissioned tailings dam, particularly for mines located in high rainfall or seismically active areas or upstream of communities.

The essential technical requirement for the implementation of submarine or deep-sea disposal of tailings is access to an offshore location with suitable oceanographic features. At some sites, this is achieved through the use of long overland and subsea pipelines. For example, the Ramu Nico Mine in Papua New Guinea transports ore via a 450-kilometre pipeline to the coast. However, most coastal locations are unsuitable for marine placement of tailings due to the shallowness of local waters, and submarine or deep-sea disposal methods are feasible for only a very few mining operations. There is also widespread opposition to DSTP. It is not permitted in some jurisdictions. As of 2020, only 15 mining operations were implementing deep sea tailings placement.

Underground mine backfill

Mining methods practised at many underground mines include placement of backfill in the mined-out voids as part of the mining cycle. The backfill provides geomechanical support to the rock mass and thereby maximises ore recovery by avoiding ore loss to non-recoverable pillars. Mine tailings, combined with a binder such as cement, are often used as the backfill. This practice can significantly reduce the quantity of tailings that must be stored on surface, but is never the complete solution. As a result of crushing and grinding, the dry bulk density of the mine tailings is always significantly lower than the in situ density of the gangue from which it was produced, so only about half of a mine's tailings can be disposed of in this manner.

CRITICAL CONTROLS

A Critical Control is a control that is essential to preventing an unwanted event or mitigating consequences should it occur. The absence or failure of a Critical Control would significantly increase the risk of an incident irrespective of the existence of other controls.

Bowtie analysis (below) is a commonly used method for understanding the causation of catastrophic events and associated Critical Controls with application to TSF failures (Mills et al. 2016). Failure Modes and Effects Analysis (FMEA) is a similar, more complex method (Santos et al. 2012).

In addition to the essential nature of Critical Controls in preventing incidents or mitigating their consequences, another test often applied to identifying these controls is that they are directly verifiable in the field or at least can be inferred from construction records. These factors set Critical Control apart from administrative or organisational risk controls. As a typical example of administrative control, safety training, while important, is never a Critical Control on the risk of accidents.

If an incident resulted from the absence or failure of a Critical Control, application of an incident investigation methodology such as Root Cause Analysis would identify this absence or failure as the immediate cause of the incident. As explained above in Why Tailings Dams Fail, the root or basic cause would most likely be organisational in nature. In the implementation of a Critical Control program, it is important not to downplay the importance of identifying and addressing the basic causes of incidents, being the ultimate reasons for the absence or failure of Critical Controls.

Implementation of a Critical Control program can support a systematic approach to risk mitigation through:

• Prioritisation of resources within the organisation for the implementation and maintenance of Critical Controls.
• Design of annual risk mitigation plans.

- Design of instrument-based condition monitoring systems to detect potentially unsafe conditions, including the implementation of trigger levels for indicating unusual or unsafe conditions.
- Design of assurance processes for risk mitigation such as inspection and audit programs.
- Implementation of key performance indicator (KPI) or "dashboard" reporting for risk management.
- Setting annual performance targets for staff involved in the management of Critical Controls.
- Ensuring workers understand the Critical Controls within their workplace.

The Critical Control approach is well established in the area of occupational safety and is being increasingly applied to the mitigation of risk associated with TSFs (Mills et al. 2016). It is applicable to any type of operational risk involving physical controls that can be verified in the field. (It may not be appropriate for all types of enterprise risk, such as financial risk.)

As one example, standard Critical Controls on the water in TSFs include freeboard allowance, spillway condition, phreatic surface within the embankment, and extended beaching with water held away from the embankment.

Information on how to implement Critical Control programs is readily available (ICMM 2015). The general approach is similar to that applied for the development of operational risk mitigation plans but with greater emphasis on workforce engagement and regular monitoring and reporting of the status of controls through inspection programs. Absent or ineffective Critical Controls identified during inspections should be managed within the site's incident management system.

CRISIS MANAGEMENT

Various tailings management guidelines, codes of practice, and standards refer to the need for a tested emergency response plan in the event of a tailings storage failure, indicating an assumption that a plan supported by drills would be sufficient to deal with the consequences of a catastrophic event. In reality, a procedure-based emergency response plan would be entirely inadequate in such circumstances.

The term "emergency response" is used to describe first-re-sponders' actions in the field, typically engaged in administering first-aid, initiating medical evacuations, and taking initial steps to render hazards safe. An effective response to a catastrophic event requires a much broader and strategic response, likely including support for teams in the field, assessment of actual and potential impacts, development and resourcing of response plans, communication with stakeholders, coordination with government disaster agencies, and dealing with the media. This level of response is described as crisis management. An effective crisis management response needs to be delivered by a well-trained crisis management team, including senior Company management. Maintenance of an effective crisis management capability by a mining Company requires a significant investment of resources, management time, consultancy fees, specialised software, and dedicated facilities. Many mining companies would likely fail this test.

RISK MANAGEMENT SYSTEMS

It has been well established that continuous improvement in the management of complex areas of risk is best achieved by implementing a management system (BP 11). Within the resources industry, the most common examples are safety and environmental management systems. Traditionally, dam safety has been viewed as mainly a technical matter, which has been judged and regulated using engineering standards. However, it has become recognised that, rather than focusing only on engineering aspects, a systematic approach based on a comprehensive management system framework that addresses human factors as well as technology is essential for minimising the risk associated with TSFs. This holds true for the management of any Catastrophic Risk associated with the resource industries.

ENGINEERING AND ADMINISTRATIVE CONTROLS

The minimisation of risk associated with TSFs relies on the application of well-established engineering and administrative controls. The following table provides a summary in this regard. More detailed guidance is available from a wide range of sources, including Elements of a Dam Safety Program (ANCOLD 2003); Guidelines on Tailings Dams – Planning, Design, Construction, Operation and Closure (ANCOLD 2012); Position Statement on Preventing Catastrophic Failure of Tailings Storage Facilities (ICMM 2016); and the Global Industry Standard on Tailings Management (Global Tailings Review 2020).

Key principles for minimising tailings disposal risk

Site selection	In the planning stage, options for deposition method and site location need to be evaluated for risk reduction opportunities. These opportunities are difficult or impossible to reclaim later in the design and operational stages. Alternative solutions for tailings disposal such as in-pit placement and DSTP as well as alternative treatment methods such as dry tailings placement should be evaluated as these may offer significant advantages in operational and closure cost and risk. The site selection process should consider factors such as Consequence Category (downstream population at risk), catchment area, foundation conditions, and storage curve.
Design process	The tailings facilities design approach and methodology should:

- Reflect industry-leading practice
- Be conservative in approach
- Conform to ICOLD or ANCOLD guidelines or equivalent
- Address all failure modes
- Utilise the most current seismic data for the site
- Take into account geochemical risk.

Design criteria	Design performance criteria should be based on Consequence Category as per ICOLD or ANCOLD Guidelines or equivalent. Performance criteria applying to Extreme Consequence Category should be adopted if the goal is to minimise tailings facilities risk (ALARP).
Closure planning	From the beginning, the design process should address the closure strategy.
Design conformance	An Engineer of Record should supervise a construction QA/QC program and monitor conformance with design. Any non-conformances should be reported to the design consultancy and site management.

(Continued)

Key principles for minimising tailings disposal risk

Change management	The design consultancy should authorise all design changes.
Tailings management	For gold mines, tailings should be treated to reduce cyanide to safe levels before placement. In the case of conventional tailings placement, subaerial deposition should be implemented with the objectives of maximising beach density and maintaining ponded water at distance from the embankment.
Water management	Operational decant volume should be minimised at all times, and the embankment lift schedule should ensure that the design freeboard allowance is available at all times.
Condition monitoring	A condition monitoring program should be in place that addresses unsafe conditions based on credible modes of failure. Trigger values for monitoring data should be established that indicate abnormal conditions and potentially unsafe conditions. Any indication of potentially unsafe conditions should be reported immediately to the design consultancy and site management.
Maintenance	Tailings facilities equipment and instrumentation should be allocated the highest priority level within the preventative maintenance system.
Competency	All staff involved in the management or supervision of tailings facilities construction and operations should possess the necessary competencies for the role.
Management of records	All records relevant to tailings facilities safety should be controlled (readily accessible and protected from loss or alteration).
Relocation of population at risk	The feasibility of relocating communities located in the zone of influence should be assessed as a means of reducing the Consequence Category and risk.
Emergency management plan	A Tailings Facilities Emergency Management Plan should be in place and regularly tested. This should provide for stakeholder engagement and include arrangements for notifying downstream communities in the event of unsafe conditions developing.

(Continued)

Key principles for minimising tailings disposal risk

Reporting	The tailings facilities design process should be fully documented in a design report to facilitate a third-party review. Regular reporting on tailings facilities risk management should be provided to senior Company management and the Board of Directors with immediate notification in the event of unsafe conditions developing.
Assurance processes	Independent expert reviews or audits of tailings facilities should be conducted regularly. Review reports and progress in addressing corrective actions arising from independent reviews should be reported to the Company management and the Board of Directors on a regular basis.
Management review	Tailings facilities management is reviewed on a regular basis by a committee comprising Company management and technical specialists with the objective of the ongoing reduction in risk in line with industry-leading practices.
Management system	The consistent application of tailings facilities risk controls should be supported by a management system (addressing targets, accountabilities, required competencies, operational controls, condition monitoring, management of corrective actions, etc.).

HUMAN FACTORS

The previous section summarises engineering and administrative controls on managing TSF risk. These are readily measurable controls that would largely be familiar to operational management teams. Also important for managing catastrophic risks are factors that may escape notice entirely, being various cognitive biases in risk perception and decision-making collectively described as human factors. It seems likely that these biases have contributed to almost all TSF failures. Yet, typically they have received limited attention by those involved in the investigation and review of these events. They include overconfidence, WYSIATI (What you see is all there is), anchoring, confirmation bias, escalating commitment, groupthink, and normalisation of deviance.

< Overconfidence – Failure to recognise and consider gaps in under-standing and the limitations inherent in engineering design processes, risk assessments, and operational controls is a common cause of bias in decision-making. It is in play when incidents at other sites are attributed to less capable management teams and when inadequate resources are allocated to implement independent reviews and other forms of assurance.

< WYSIATI – This bias concerns the tendency to form impressions and make judgements based on the information we have available, rather than question if information is sufficient to represent the situation accurately (Kahneman, 2012).

< Anchoring – Anchoring describes a situation where decision-making is based on information or assumptions treated as absolute and, accordingly, never revisited. Failure to update risk assessments or operational controls is a common example of anchoring.

< Confirmation Bias – Confirmation bias results when an individual or group favours information that affirms prior decision-making or established practices. The discounting of monitoring non- conformances as special cases and excluding data points from statistical analysis based on being outliers are common examples.

< Escalating Commitment – In the face of increasingly poor performance, individuals and groups tend to escalate commitment to an existing course of action rather than present an error to the organisation and write-off losses. An example could be a continuation of the design process for a TSF at a site after data is obtained, indicating weak foundation conditions instead of revisiting the site selection process and starting over at a new location.

< Groupthink – Groupthink refers to a bias in decision-making driven by the desire for harmony, conformity, and avoidance of conflict with others. It is commonly exhibited during workshops or meetings where junior team members are reluctant to challenge more senior members' opinions. This is the reason why "anonymous" electronic voting systems are useful for enterprise risk assessment workshops, particularly those attended by Company executives.

< Normalisation of Deviance – Normalisation of deviance occurs when departures from desirable conditions or

non-conformances become tolerated due to repetition and or familiarity, reducing the perception of risk. Management of TSF risk is particularly susceptible to this bias, as the probability of failure following a single non-conformance in condition monitoring data may be very low. The bias can be controlled by embedding risk-based thinking within a Company's culture, training employees to ensure they understand the significance of non-conformances, formalising incident reporting requirements, and implementing systems that deliver automated alerts when exceedances are recorded by instrumentation.

To conclude, for some sites, there is no solution available for low-risk tailings disposal. Mining companies need to consider the long-term risk associated with tailings disposal, including site factors such as seismicity and foundation conditions. This will inevitably mean some mineral deposits should not become mines.

Acronyms and abbreviations

ACM	Asbestos-Containing Material
ADB	Asian Development Bank
ALARP	As Low as Reasonably Practicable
ARD	Acid Rock Drainage
AS	Ambient Standard
ASX	Australian Securities Exchange
BP	Boardroom Perspective
CBD	UN Convention on Biological Diversity
CEC	Contaminant of Emerging Concern
CERCLA	Comprehensive Environmental Response, Compensation, and Liability Act
CMP	Crisis Management Plan
CMR	Crisis Management Room
CMT	Crisis Management Team
CoP	Cessation of Production
CR	Critically Endangered
CSR	Corporate Social Responsibility
DD	Data Deficient
DSTP	Deep-sea Tailings Placement
EDC	Endocrine-disrupting Chemical
EHS	Environment, Health, and Safety
EIA	Environmental Impact Assessment
EN	Endangered
EP	Equator Principles
ES	Effluent Standard
ESA	Environmental Site Assessment
ESDD	Environmental and Social Due Diligence
EPFI	Equator Principles Financial Institution
ESG	Environmental, Social, and Governance

ESIA	Environmental and Social Impact Assessment
ESMS	Environmental and Social Management System
ESV	Enlightened Shareholder Value
EW	Extinct in the Wild
EX	Extinct
FIFO	Fly-in, fly-out
FMEA	Failure Modes and Effects Analysis
FPIC	Free Prior Informed Consent
GHG	Greenhouse Gas
GIIP	Good International Industry Practice
GISTM	Global Industry Standard on Tailings Management
GRI	Global Reporting Initiative
GRTK	Genetic Resources related to Traditional Knowledge
ICMM	International Council on Mining and Metals
IED	Industrial Emissions Directive
IFC	International Finance Corporation
IFI	International Financial Institution
ILO	International Labour Organisation
IoT	Internet of Things
IPCC	Intergovernmental Panel on Climate Change
IUCN	International Union for Conservation of Nature
KPI	Key Performance Indicator
LARAP	Land Acquisition and Resettlement Action Plan
LC	Least Concern
LTIFR	Lost Time Injury Frequency Rate
LTO	Licence to Operate
M&A	Merger and Acquisition
MMSD	Mining, Minerals and Sustainable Development
NG	Net Gain
NGO	Non-governmental Organisation
NNL	No Net Loss
NORM	Naturally Occurring Radioactive Material
NT	Near Threatened
OECD	Organisation for Economic Co-operation and Development
OHS	Occupational Health and Safety
PCB	Polychlorinated Biphenyl
PDCA	Plan-Do-Check-Act
PPE	Personal Protective Equipment
PPP	Polluter Pays Principle
PRI	Principles for Responsible Investment

PS	Performance Standard
RAP	Resettlement Action Plan
RCRA	Resource Conservation and Recovery Act
RP	Resettlement Plan
SLO	Social Licence to Operate
SOP	Standard Operating Procedure
SWOT	Strengths, Weaknesses, Opportunities, Threats
TBL	Triple Bottom Line
TIFR	Total Injury Frequency Rate
TRIPs	Trade-Related Aspects of Intellectual Property Rights
TSF	Tailings Storage Facility
UNDRIP	United Nations Declaration on the Rights of Indigenous Peoples
UNEP	United Nations Environment Programme
UNPRI	United Nations-supported Principles for Responsible Investment
VU	Vulnerable
WGAC	Wintawati Guruma Aboriginal Corporation
WRG	Water Resources Group
WTO	World Trade Organization
WYSIATI	What You See Is All There Is
3BL	Triple Bottom Line
3Ps	Profit, People, and Planet

References

ANCOLD. (2003). Guidelines on dam safety management. https://www.ancold.org.au/.

ANCOLD. (2012). Guidelines on tailings dams, planning, design, construction, operation and closure. https://www.ancold.org.au/.

ASX Corporate Governance Council. (2014). *Corporate Governance Principles and Recommendations*, 3rd ed. www.asx.com.au.

Atkisson, A. (2012). *Life Beyond Growth. Annual Survey Report*, (Institute for Studies in Happiness, Economy, and Society, Tokyo, Japan).

Australian Government (2016). *Mine Closure. Leading Practice Sustainable development Program for the Mining Industry*. September 2016

Barbier, E.B. (1987). The concept of sustainable economic development, *Environmental. Conservation*, Vol. 14, No. 2; pp. 101–110.

Barizah, N. (2020). Indonesia's patent policy on the protection of genetic resources related to traditional knowledge: Is it a synergy to fulfill the TRIPs agreement and CBD compliance? Yuridika, *E-Journal of Faculty of Law, Universitas Airlangga*, Vol. 35, No. 2; pp. 321–342.

Chambers, D.M. and Higman, B. (2011). Long term risks of tailings dam failure. www.csp2.org.

Cone. (2016). Three-quarters of millennials would take a pay cut to work for a socially responsible company. http://www.conecomm.com.

COSO. (2017). Enterprise risk management - integrating with strategy and performance. *Committee of Sponsoring Organizations of the Treadway Commission*. www.coso.org.

Davies, M.P. (2002, September). Tailings impoundment failures: are geotechnical engineers listening. *Geotechnical News*.

DFAT. (2016). *Leading Practice Handbook: Mine Closure. Leading Practice Sustainable Development Program in the Mining Industry*. Industry. gov.au - DFAT.gov.au.

Dobson, A. (1996). Environment sustainabilities: an analysis and a typology, *Environmental Politics*, Vol. 5, No. 3; pp. 401–428.

Duijm, N.J. (2015). Recommendations on the use and design of risk matrices, *Safety Science*, Vol. 76, pp. 21–31. Doi: 10.1016/j ssci.2015.02.014.

Elkington, J. (1994). Towards the sustainable corporation: win-win-win business strategies for sustainable development, *California Management Review*, Vol. 36, No. 2; pp. 90–100.

Elkington, J. (2018, June 25). 25 years ago I coined the phrase "Triple Bottom Line." here's why it's time to rethink it, *Harvard Business Review*, Vol. 25; pp. 2–5.

Ferguson, K.I. (2015). A study of safety leadership and safety governance for board members and senior executives. PhD thesis, Queensland University of Technology.

Fink, L. (2020). A fundamental reshaping of finance. https://www.blackrock.com/corporate/investor-relations/larry-fink-ceo-letter.

GISTM. (2020). Global industry standard on tailings management. GlobalTailingsReview.org.

Greencorp. (2020). Land disturbance mapping 2000 to 2020; confidential mining client.

Hoekstra, A.Y. and Mekonnen, M.M. (2012). The water footprint of humanity, *Proceedings of the National Academy of Sciences*, Vol. 109, No. 9; pp. 3232–3237.

Holmberg, J., Bass, S., and Timberlake, L. (1991). *Defending the Future: A Guide to Sustainable Development*, (IIED/Earthscan, London).

ICMM. (2015). *Critical Control Management Implementation Guide*. www.icmm.com.

ICMM. (2019). *Integrated Mine Closure: Good Practice Guide*, 2nd ed. www.icmm.com.

IFC. (2002). *Handbook for Preparing a Resettlement Action Plan*. www.ifc.org.

IRM. (2018). *A Risk Practitioners Guide to ISO 31000: 2018*. www.theirm.org.

Jaques, T. (2007, June). Issue management and crisis management: an integrated, non-linear, relational construct, *Public Relations Review*, Vol. 33; pp. 147–157.

Kahneman, D. (2012). *Thinking Fast and Slow*, (Penguin Books). ISBN: 9780141033570.

Lawson, C. and Adhikari, K. (2018). *Biodiversity, Genetic Resources and Intellectual Property: Developments in Access and Benefit Sharing of Genetic Resources*, (Taylor & Francis Group, Routledge).

Locke, R. (2003). The promise and perils of globalisation: the case of Nike, In Kochan, T.A. and Schmalensee, R.L. ed., *Management: Inventing and Delivering Its Future*, (MIT Press, Cambridge, MA).

Mekonnen, M.M., Gerbens-Leenes, P.W., and Hoekstra, A.Y. (2015). The consumptive water footprint of electricity and heat: A global assessment, *Environmental Science: Water Research & Technology*, Vol. 1, No. 3; pp. 285–297.

Mills, R., Rebecca, F., and Barker, M. (2016). Would bowties and critical controls contribute to the prevention of high consequence /low frequency dam failures? ANCOLD.

MMSD. (2002). *Breaking New Ground: Mining, Minerals, and Sustainable Development (MMSD) Project*. Published by Earthscan for IIED and WBCSD. www.iied.org › mmsd-final-report.

Morello, E., Haywood, M., Brewer, D., and Asmund, G. (2016). The Ecological Impacts of Submarine Tailings Placement. In Hughes, R. N., Hughes, D. J., Smith, I. P., and Dale, A. C. (Eds). *Oceanography and Marine Biology: An Annual Review*, (Taylor & Francis), pp. 315−366.

Owen, J.R., Kemp, D., Lebre, E. Svobodova, K., and Perez Murillo, G. (2020). Catastrophic tailings dam failures and disaster risk disclosure, *International Journal of Disaster Risk Reduction*, Vol. 42; p. 101361.

Parker, L. (2018, October 22). In a first, microplastics found in human poop, *The National Geographic Society*. www.nationalgeographic.com/ environment/article/news-plastics-microplastics-human-feces.

PwC. (2019). Global crisis survey. www.pwc.com.

Robinson, J. and Tinker, J. (1998). Reconciling ecological, economic, and social imperatives. In Schnurr, J. and Holtz, S. (Eds). *The Cornerstone of Development: Integrating Environmental, Social and Economic Policies*, (International Development Research Centre, Ottawa), pp. 9–44.

Rolsky, C. (2019, June 6). The vertical distribution and biological transport of marine microplastics across the epipelagic and mesopelagic water column, *Environmental Science and Technology*.

Santamarina, J.C., Torres-Cruz, L.A., and Bachus, R.C. (2019, May 10). Why coal ash and tailings dam disasters occur, *Science*, Vol. 364, No. 6440; pp. 526–528.

Santos, R.N.C.D., Caldeira, L.M.M.S., and Serra, J.P.B. (2012). FMEA of a tailings dam, *Article in Georisk Assessment and Management of Risk for Engineered Systems and Geohazards*, Vol. 6, No. 2; pp. 89–104.

Spiliakos, A. (2018, October 10). What Is Sustainability in Business? | HBS Online. Business Insights - Blog. https://online.hbs.edu/blog/post/ what-is-sustainability-in-business.

Spitz, K. and Trudinger, J. (2019). *Mining and the Environment – From Ore to Metal*, 2nd ed., (Taylor & Francis Group, London, UK).

Tague, N.R. (2005). *Plan–Do–Study–Act Cycle. The Quality Toolbox*, 2nd ed., (ASQ Quality Press, Milwaukee) pp. 390–392. ISBN 978-0873896399.

Tricker, B. (2015). *Corporate Governance: Principles, Policies, and Practices*, 3rd ed., (Oxford University Press, Oxford, UK).

WCED. (1987). *Our Common Future. World Commission on Environment and Development* (also referred to as the Brundtland Report). www.sustainabledevelopment.un.org.

World Economic Forum. (2018). *The Global Risks Report 2018*. www.weforum.org.

WRG. (2019). *2019 Annual Report. The 2030 Water Resources Group*. www.2030wrg.org.

Index

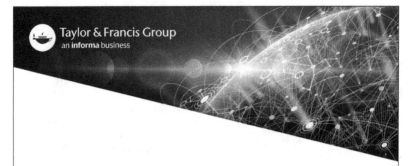

CPSIA information can be obtained
at www.ICGtesting.com
Printed in the USA
JSHW022033090522
25519JS00001B/109